Crafts studies
Series editor: N. Hiller

Electronic systems for mechanics

Volume 1

D.H.E. King
Lecturer in Electronics
West Suffolk College of Further Education
Out Risbeygate
Bury St. Edmunds
Suffolk, England

Peter Peregrinus Ltd.

Published by Peter Peregrinus Ltd.
Southgate House, Stevenage,
Herts. SG1 1HQ, England

First published 1974

ISBN: 0 901223 58 1

Set in cold type by EWC Wilkins Ltd, London and
Printed in England by Billing & Sons Ltd., Guildford and London

This is the first of two volumes that have been written for Radio, Television and Electronics Mechanics. They are based on the Electronic Systems syllabuses of the City & Guilds of London Institute (CGLI) Course 222, Parts I and II. The material in this first volume deals mainly with the Part I syllabus, and the second volume deals with the Part II syllabus. It is inevitable that the syllabuses are not covered exactly, in detail, in the two volumes, since this would entail considerable repetition and there are obvious natural limitations on the contents. But the essential details of the building blocks of electronic circuits or systems are explained in such a way that any mechanic undergoing training or employed in the radio, television or electronics industry can develop a sound understanding of the functions performed by these blocks.

The mathematics has been kept to a minimum so that the mechanic may read through any section of the text without hesitation, and a parallel knowledge of the attendant electrical principles has been assumed.

The keynote to these volumes is simplicity, and the language has been chosen so that the mechanic may feel that he is on familiar ground when reading the descriptions of particular systems or circuits.

Bury St. Edmunds D.H.E.King

Contents

Contents of chapters

Chapter 1

Safety

1.1 GENERAL

The following Sections may seem too gloomy and might suggest that unconscious mechanics are littered around the world owing to the effects of electricity or electronics. This is not so, of course, and yet the chance of danger both to the mechanic and to the user or customer from the apparatus is very high IF precautions are ignored. The user is expected to be ordinarily careful and not to drop his toaster or television set in the bathtub, but the designer and the mechanic must realise the other possible dangers and ensure the safety of everyone (including themselves) by taking the correct precautions at all times.

1.2 MAINS SUPPLIES

Whether or not we know how electrical apparatus works, we invariably use the mains electricity supply to power the apparatus for long periods. Although correctly designed apparatus is safe to use, a mechanic spends much of his time inside the 'works' and must be familiar with working at the high voltages.

A two-wire mains supply usually has one wire earthed at the supply authority's substation; this wire is called the NEUTRAL. The other wire is normally at 240 volts with respect to the neutral and is the LIVE wire. If three wires or a 3-pin socket is used, the third wire will be the true EARTH connection to a nearby metal cold-water pipe or an earth-stake, as in Fig. 1.1.

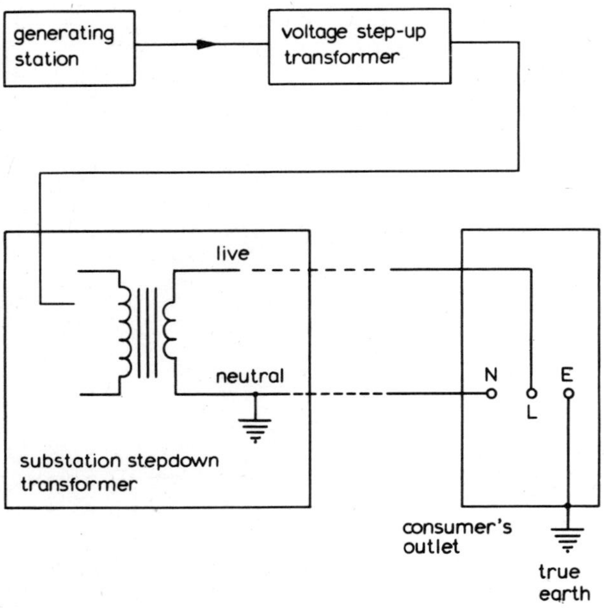

Fig. 1.1 Mains electricity supply system

The power input to apparatus is supplied by NEUTRAL and LIVE and there should be no direct connection between either of these wires and the apparatus casing. The EARTH wire should be connected to any exposed metal parts, casing or chassis, and this will ensure that the user is safeguarded against any leak that may occur from either side of the mains supply; the leak would be bypassed to earth directly rather than through the user's body.

Care must be taken with the correct connections of colour-coded mains wiring to plugs. The present British Standard colours are:

1973 standards		**'Old' colours**
Live	Brown	Red
Neutral . . .	Blue	Black
Earth	Green/Yellow striped	Green

while the 'older' colours, still used with much apparatus, are as shown. It is not uncommon for a home-fitted plug to be wired wrongly through ignorance, or for imported apparatus to be fitted with nonstandard colours. Check for correct earthing and colour-coding when repairing or installing apparatus, and also check for correct fuse-ratings of the fuses fitted in British 13 A plugs; a fuse that is rated at 13 A for a load only needing perhaps 0·5 A could lead to very hot wiring between the plug and the load before the too-highly rated fuse finally blew. Fig. 1.2 shows the inside view of a British 13 A plug.

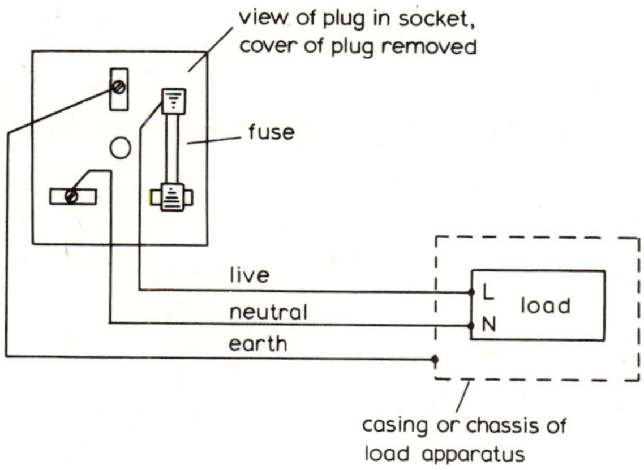

Fig. 1.2 13 A plug wired to load

1.3 ELECTRIC SHOCK

There is a saying:

'It's the volts that jolts
But the mills that kills'

and this refers to the initial shock that is caused by a high voltage and the effect if current continues to flow for only a few seconds.

Approximate maximum safe values are 50 V or 5 mA but many people can feel and would be shocked by considerably less than these values.

It must be emphasised that even quite a small shock can be dangerous; PRACTICAL JOKES CAN KILL. You cannot afford to take chances with sloppy connections, careless wiring or a slapdash attitude that may blow a fuse – or a life.

A person with damp hands or one who is standing on a damp surface is very liable to feel the shock before someone else who is drier; particularly in cellars, laundries, lavatories, bathrooms or outside when standing on the ground. Apparatus and wiring used in such circumstances must be in as perfect a condition as the mechanic (or user) can make it. Pay special attention to correct earthing of apparatus.

1.4 SHOCK TREATMENT

If someone is badly shocked into unconsciousness and you are present, then obviously you need to turn off the supply. A distant switch may take some time to reach, so disconnect the disabled person in this instance by immediately moving him with dry wood or pulling him by his clothing. Once he is safely removed from the cause of the shock, listen for his breathing; if none, start an artificial respiration procedure at once and shout for help.

Artificial respiration to a person who is not breathing should be continued for not less than an hour, or until one is told by a doctor to stop. Learn how to give the 'kiss of life'; it may save a life and there are only about four minutes after the onset of breathing failure in which to start helping if permanent brain damage is to be avoided.

If the shocked person is only unconscious, place him on his side, make sure that he will remain warm and obtain help. Do NOT allow any alcohol to be given, whether asked for or not.

1.5 APPARATUS INPUT WIRING

Fig. 1.3 shows the preferred and safest input circuitry with a switch in the LIVE lead, followed by a correctly rated FUSE and then by a PILOT LAMP, preferably a neon type that is less liable to fail than one with a filament. The lamp thus indicates correct operation of both the on/off switch and the fuse.

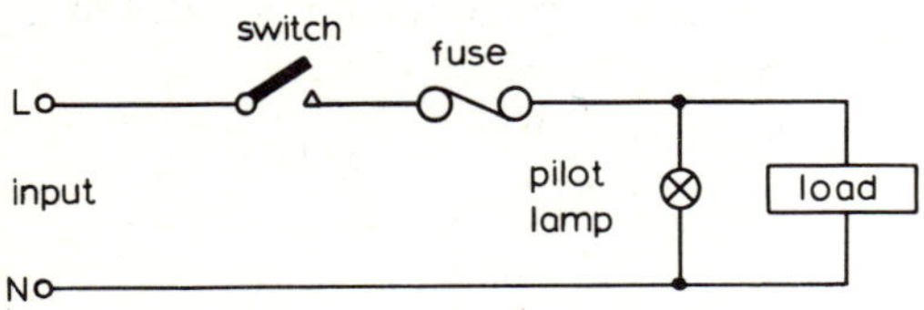

Fig. 1.3 Input circuit to load

Always bear in mind, however, that any component is liable to be faulty – even brand-new, just-unpacked components – so while a faulty switch ought to be indicated by the pilot lamp, a faulty lamp could suggest that the fuse is blown, or the switch open. Thus even an apparently obvious set of symptoms like

switch on . . . .
pilot lamp still dark . . . .
therefore fuse or supply lead is open

must be considered carefully. There is the possibility that the obvious fault is not the correct one, and there could be danger present for the mechanic.

The only absolutely safe isolation of apparatus is obtained by removing the mains plug from its socket, but even after isolating from the mains it is still possible for charged capacitors within the apparatus to retain their charge for days and to give a shock.

1.6 THE SERVICE MECHANIC

The service mechanic will often have to work upon 'live' equipment – switched on and running so that the fault symptoms show up – for test gear to be used and for the apparatus to be finally adjusted to the user's requirements. The list of do's and don'ts to be followed cannot be exact since the apparatus, the repair facilities and the test gear used will vary enormously.

For each fresh installation or repair it is again emphasised that you need to check for:

correct plug or socket wiring to the supply (mains or battery)
correctly colour-coded supply wiring
correct rating(s) of the fuses used
correct operation of the on/off switch and pilot lamp (if fitted)

If not done at the beginning, these will need to be done anyway before returning the equipment as 'serviceable', and these checks will give the mechanic confidence while he is dealing with the particular equipment.

Then, if the circuit must be investigated, first switch off and remove the plug from the supply. Then LOOK for faulty components or wiring using your eyes only – if you must move anything, use one hand only and do not hold the chassis with the other hand. In HM Forces this is the only time that a man is told to keep one hand in his pocket! An earthing wire with one end clipped to the chassis and the other end dabbed against exposed (possibly charged) terminals will remove safely any stored charge via the wire rather than through the hand or body.

Having seen or smelt the fault, it may be attended to; if not located by now the fault must be traced with the equipment switched on, and even greater care must be taken.

Use isolating transformers (Fig. 1.4) for feeding direct-connection chassis as are found in most t.v. receivers in Britain; handle the apparatus as little as possible whilst it is switched on (and with one hand, remember) and make test-gear connections with care.

After curing the fault and replacing the chassis in its case or cabinet, make a final check that no live parts may be easily touched by a non-technical person – including children crawling around the back of a radio, t.v. receiver or washing-machine. Apparatus installed in factories is more liable to be tampered with by adults, but it is still the responsibility of the mechanic to anticipate either age and to safeguard both.

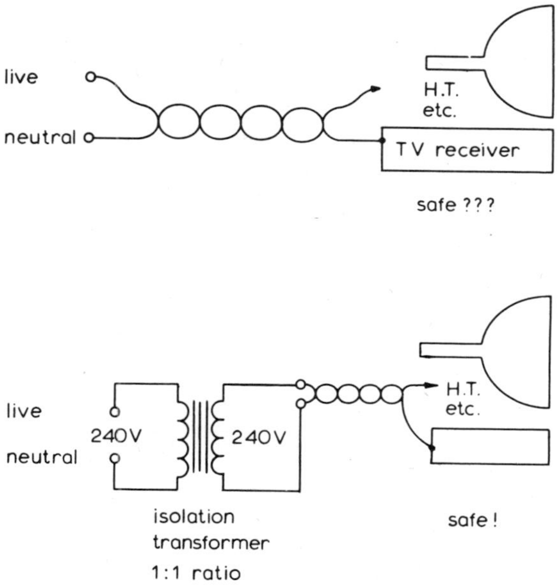

Fig. 1.4 Need for isolation transformer to give safety

As a mechanic, be proud to have left equipment in a good working condition. Tell the user of any dangers if you are not sure of absolute safety; your responsibility does not end when you walk away from the job.

1.7 ELECTRICAL FIRES

Within small apparatus fires are simply dealt with by switching off, but unless the supply lead is removed there is still the normal shock risk if conducting liquids contact the live wiring. Within large or complex installations the isolation switching may not be readily accessible, and a fire-depressing gas like carbon dioxide, a heavy-gas-producing liquid like carbon tetrachloride (C.T.C.), or an asbestos blanket must be used rather than water or foam which are electrical conductors. These heavy gases extinguish the fire by excluding oxygen; humans also need oxygen to breathe, so take care not to remain in the room to suffocate as well. In addition C.T.C., which is often used for cleaning and degreasing, produces a poisonous gas if inhaled through a cigarette and has resulted in several deaths. In general, all volatile liquids should be used only in an adequately ventilated room.

1.8 SOME SPECIAL RISKS

Old and discarded t.v. fluorescent tubes are coated inside with a powder to help produce light when in use. When the tube is broken this powder will fly around in a cloud, and it is POISONOUS; pack the tubes in labelled boxes from which the replacement tubes have been removed, and allow the refuse collector to deal with the old tubes; the collector will know the best method of disposal.

Any t.v. tube has a high air pressure upon the outside surface and small scratches weaken the surface to the point where implosion may occur (the glass is pushed in by the air rather than out as with an explosion). The tube having imploded, the glass keeps travelling and the sight of a sideboard with 12 mm thick glass embedded in the wood (from a wrecked t.v. set three metres away) has impressed on the author the need to handle and dispose of these tubes very carefully. Luckily, this incident occurred in the middle of the night; a faulty on/off switch was the cause of a small flare-up just beneath the tube, and the resulting blast from the imploding tube extinguished the fire.

Some specialised tubes, valves and transistors, while quite safe to handle and use, are made with very poisonous materials inside; they must not be dismantled or sawn apart to 'look at the insides'. The manufacturer often attaches a label asking for these components to be returned when faulty, for their safe disposal.

Colour t.v. receivers emit X-rays from the e.h.t. rectifier compartment and the t.v. tube. Sufficient shielding and materials will have been fitted to ensure safety to the viewer at the normal viewing distances. Closer to these radiating components then the viewer, the mechanic runs little risk unless working with this small portion of the receiver day after day. Leave the shielding in place when possible and do not regard it as just another spare bit of metal fitted to make the job more awkward.

Every job has its own hazards, physical or electrical. It is not 'clever' or 'with it' to disregard these hazards and to think that you are immune. The fussy old chargehand or boss has managed to grow old by taking care and being fussy – he can remember those that did not take care.

1.9 TEST GEAR AND TOOLS

There should be little difference in the mechanic's attitude to tools and test gear; both are used to investigate and repair apparatus, and it should be the mechanic's aim to own as much of his own gear as possible and to take pride in its condition. Without being selfish, it is recommended that as little as possible is lent to others – and, of course, that means that your mates will not lend to you – because only the owner will really look after the tools. This rule cannot apply to the expensive, specialised or uncommon item, of course, so the other rule to follow is that you return the borrowed item as soon as it is finished with, even if inconvenient to you. This will keep a clear conscience, a clear bench and your friends.

Chapter 2

Unidirectional supplies and signalling

2.1 ELECTRICAL UNITS

Electrical principles teach us that electricity is to do with electrons. A current is a flow or movement of electrons which are so small that they are able to move around in some gases, liquids and solids with little hindrance and yet be kept in place by surrounding the conducting material with a nonconducting or insulating material.

About 200 years ago came the beginning of 'current electricity' with A. Volta in Italy, who found that two different metals in a suitable liquid gave him a source of electricity that was available continuously. Up to this time only static or 'charges' of electricity had been used, lightning being the main source at very high pressures and, of course, this was very dangerous. Rubbing amber or glass with silken cloths also produced this static electricity, and ingenious machines were invented to use frictional methods to produce very high pressures – long sparks – but very small continuous flows.

Volta's discovery produced very small pressures but *did* enable experiments to be carried out with the continuous flow of current. The unit of electrical pressure is of course the VOLT, and a French inventor called Ampére is commemorated in the unit of current because of his investigations into the measurement of the amount of flow rather than just detecting the presence of a current in a conductor.

Nowadays we are quite used to talking about voltage, of current measured in amperes and using conductors of copper wire already insulated and in 100 m or 1000 m lengths. It is interesting, however, to think back only a short while to the time when electronics and electricity did not exist as trades, only as experimental hobbies for men with enquiring minds and enough money to afford the peculiar materials needed.

Direct current, usually abbreviated and written as d.c., was mostly supplied from dry cells (the inventor being Lechanché of France) or from series- and parallel-connected cells, which are then termed 'batteries'. Single cells produce voltages of about one to two volts, while batteries may be made up to produce hundreds or even thousands of volts, and nowadays there are many different types of cell and battery manufactured with assorted prices, advantages and sizes. Power supplies are also available to produce d.c. from the electrical mains supply, in place of cells or batteries of cells.

2.2 INFORMATION AND SIGNALS

The words that you are reading are information, produced with black wiggly lines on paper by a printer and then sent to the reader who uses light, sight and his brain to interpret the letters and words as information.

Spoken words are produced by the mouth and the throat varying the air pressure. These variations travel to the listener's ear and his hearing and brain interpret the sounds as information.

A blind and deaf person has to rely on his sense of feeling or touch, his sense of smell, and lastly his sense of taste. So far taste has been little employed for signalling purposes, but Braille printing methods require the sense of touch, and insects and moths use the sense of smell to attract mates over a distance of many thousand metres. A human's range of smell is rather more limited!

Fig. 2.1 shows how INFORMATION is transferred from one place to another. The information may need to be pruned to the minimum, like yes/no, go/stop, do/do not, 0–1, or, like a telegram, be limited by the system cost.

The MODULATION system or method that is chosen adapts the information to fit cost, distance and transmission needs. To a person across the room one speaks via air-pressure variations. To the next town a written letter may be sent. To an oncoming line of traffic there are stop/go lights signalling to the driver's eyes, while far-distant places use radio for fast communication together with specialised modulation systems.

A TRANSDUCER is the name given to a device that converts one form of energy to another. A microphone, loudspeaker, petrol engine, soldering iron, the whole of our bodies and five senses; all are transducers. The information in Fig. 2.1 is now adapted or modified by the chosen transducer to fit the transmitter, which then

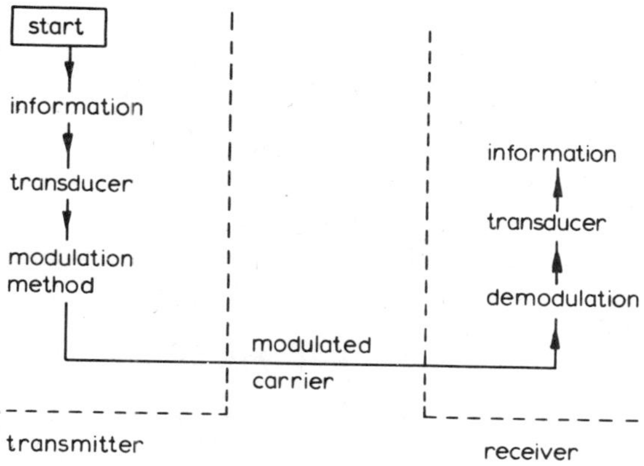

Fig. 2.1 Transfer of information

transmits the modulated carrier.

The CARRIER system may be air	....	sound
paper	....	written letters and books
cables	....	telegraphy and telephony
electromagnetic radiation	....	radio, t.v. and sight

It is perhaps surprising that sight uses the same carrier system as radio but this is described further in Chapter 4.

At the RECEIVER the information has to be removed from the carrier by a DEMODULATION process. In other words, the message must be deciphered from the particular method used to convey it, and this will require that the receiver 'knows' about the modulation method.

Finally, another transducer converts the information into a form suitable for use, usually by the brain. If a person is controlling a model aircraft by radio, then of course the final purpose of the transmitted information is to steer the aircraft around the sky, but it could be argued that there is retransmission of information about the aircraft movement back via light and sight to the operator's brain. This would be called FEEDBACK and will be discussed later in more detail.

2.3 SIGNALLING WITH D.C.

From about 1750 to 1850, inventors concentrated on sending messages by wire-cable telegraphy – and the names of Morse, Wheatstone, Bell and Edison are famous. These men lived when engineering and materials had become available in sufficient quantity to allow their inventions to be made use of cheaply, but the sending of information was still limited by the relatively crude methods then available.

A d.c. signal can be turned on and off by a switch; lamp signals are of limited range, even in the dark, but using long pairs of wires with a receiving electromagnetic sounder permitted ranges of thousands of metres to be achieved. It was the modulation or coding of the information that proved difficult to achieve, and Samuel Morse had the idea of representing letters, numerals and punctuation marks by a series of dots and dashes – the Morse Code – that is still in use today. Some of the letters are given in Fig. 2.2, where a message is spelt out in code and a typical signalling circuit is shown.

The Morse Code is an adaptation of stop/start signalling, but it suffers from the disadvantage that a very weak signal gives an uncertain deflection of the sounder or pen. Even with very sensitive intermediate relays, the difference between a small current and zero current is uncertain, so a modification to use current reversal was introduced. If the current is actually too small to be detected, this method fails as well, but uncertain currents give better results with the circuit in Fig. 2.3.

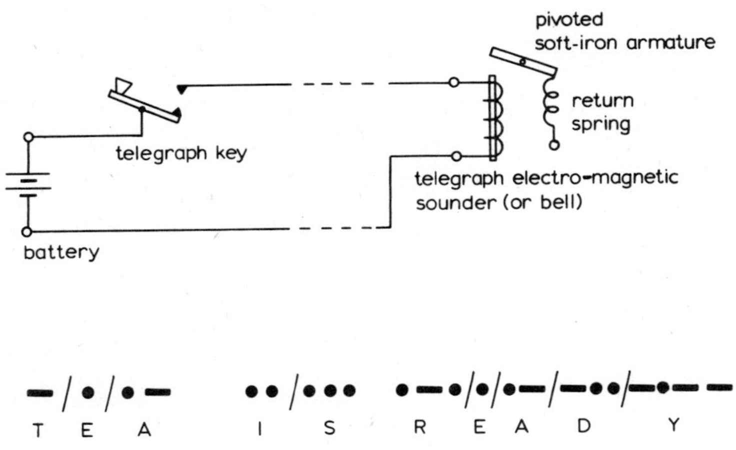

Fig. 2.2 Telegraphy circuit and Morse message

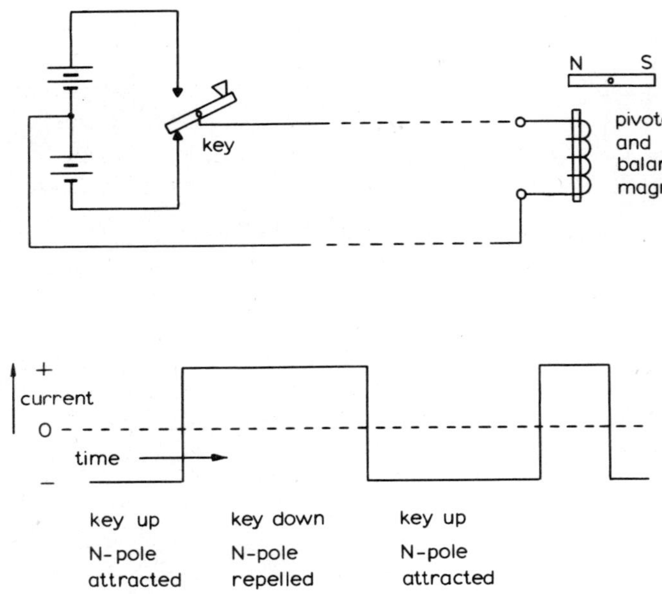

Fig. 2.3 Reversing-current circuit and waveform

Mechanical decoding of the Morse Code's varying quantities of dots and dashes per symbol is difficult, so the Murray Code was invented for machine use. Five signals, called MARKS and SPACES, and involving current reversal, are used for each symbol, and these are conveniently sent by the circuit in Fig. 2.3.

The decoding apparatus may also be made to respond only to 'bundles of five', and if necessary to request re-transmission of a poorly received message – all without the operator's help. Fig. 2.4 *a* shows a portion of the Murray Code as sent with reversing-polarity currents. The sequence of pulses commands the teleprinter to start with a 'space' signal and finishes the letter or symbol with a 'mark' signal that is $1\frac{1}{2}$ units of time. Fig. 2.4*b* shows a section of perforated tape with code 'marks' punched in; read the code downwards, 'A' being mark-mark-space-space-space, with a row of paper-tape drive holes between the second and third code symbols.

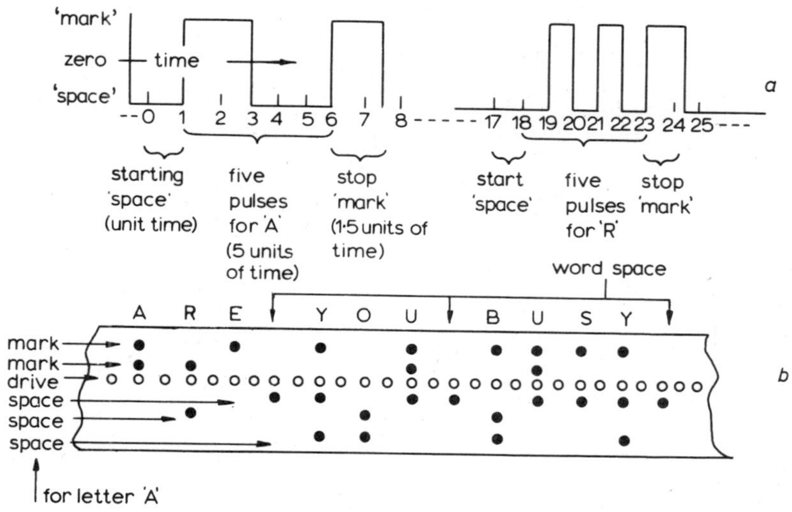

Fig. 2.4 (*a*) Teleprinters Murray Code
(*b*) punched-tape message

Motors may be connected to the on/off d.c. instead of sounders or lamps, but with the difference that the motor might need to run at intermediate speeds between full speed and zero. Once again a code method may be used; the d.c. is switched on-off-on-off sufficiently fast for the motor armature's flywheel effect to keep it turning continuously, but the ratio of on:off is chosen so that the resulting speed is as required. The graph in Fig. 2.5 is

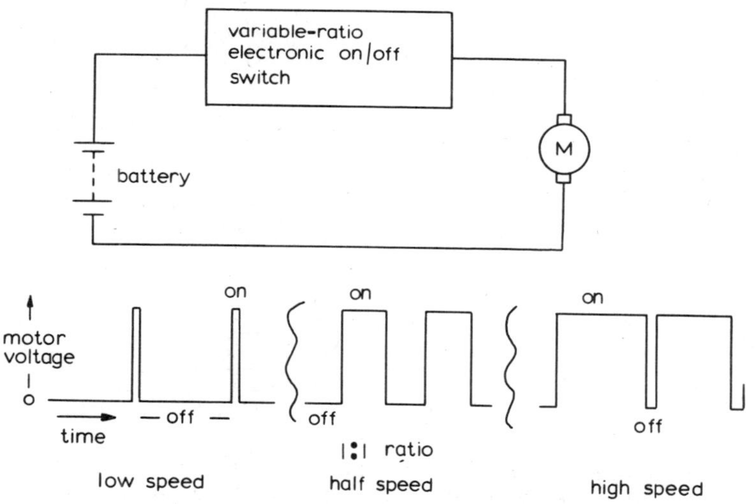

Fig. 2.5 Motor control by variable mark/space ratio d.c.

to a base of time, and if ON is called a MARK while OFF is called a SPACE, then with a high mark/space ratio the motor speed is high. A low mark/space ratio gives low speeds and a 1:1 mark/space ratio gives a half-speed effect. The fast switching requires electronic control, but the circuit is fairly simple and gives far better control of model trains, cars and electric drills than does the older method of inserting a series resistor or rheostat. This 'proportional pulse' method has the great advantage that far less power is wasted in the control system when the motor is running at low speeds.

Chapter 3

Alternating supplies and signalling

3.1 ALTERNATING CURRENT

If a voltage or current is continually being reversed so that the polarity or flow is continually changing direction or alternating, then we talk about an ALTERNATING CURRENT or a.c. supply. Strictly speaking, there cannot be 'a.c. voltage', and yet this loose way of referring to an alternating voltage is quite often used. There is a difference between an alternating supply and the on-off-on-off supply considered in the previous chapter, since there the direction of flow was always in the same direction while an a.c. supply must regularly change direction.

An ordinary centre-zero d.c. instrument could only be used to measure the size of the changing voltage or current if the rate-of-change were sufficiently slow for the instrument movement and pointer to respond, i.e. only up to once or twice per second. As the rate-of-change is increased the movement is unable to follow the fast changing values, so it indicates zero, having insufficient time to move to a positive reading before the voltage changes to a negative value, and so on.

3.2 WAVEFORM

An oscilloscope (or scope) trace is able to move very fast indeed and the fluorescent spot is easily able to follow alternating changes of up to a million per second, or even faster. The vertical deflection on the cathode-ray tube (or c.r.t.) indicates the amplitude in conjunction with the range or sensitivity setting, while the 'scope timebase' moves the trace horizontally over a period of time. The resulting waveform shows how the voltage (or current) varies above and below zero, indicating not only the size or amplitude in the vertical direction but also the shape of the a.c. generator waveform.

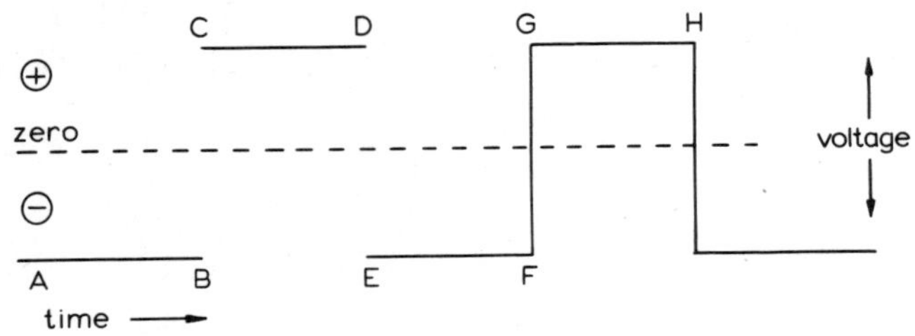

Fig. 3.1 Square waveform

Fig. 3.1 is the voltage waveform resulting from a changeover switch such as was suggested in Chapter 2 as a modification of d.c. signalling. There is no trace shown between *B* and *C* or between *D* and *E*, suggesting that the changeover of polarity is infinitely fast and that the switch can snap from one position to the other in zero time. This is of course impossible, and the more realistic waveform is continued EFGH where the verticals will be just visible on a scope.

By using a variable resistor connected as in Fig. 3.2*a*, variation of the sliding contact position along the resistor track allows the waveform in Fig. 3.2*b* and *c* to be obtained, the difference being that in Fig. 3.2*c* there is no resting period at either end of the track before reversing the direction of movement.

Of the many voltage shapes or waveforms that could be produced by moving the slide to either side of the centre, a very special shape is that of Fig. 3.2*d*, a pure SINE WAVE. This waveform could also be produced by rotating a coil in a magnetic field, using slip-rings to pick off the voltage from the coil, or by using an electronic generator. The sinusoidal waveform or sine wave is a naturally occurring shape in nature and is usually taken as a basic reference waveform, all other periodic waveforms being a combination of various sine waves.

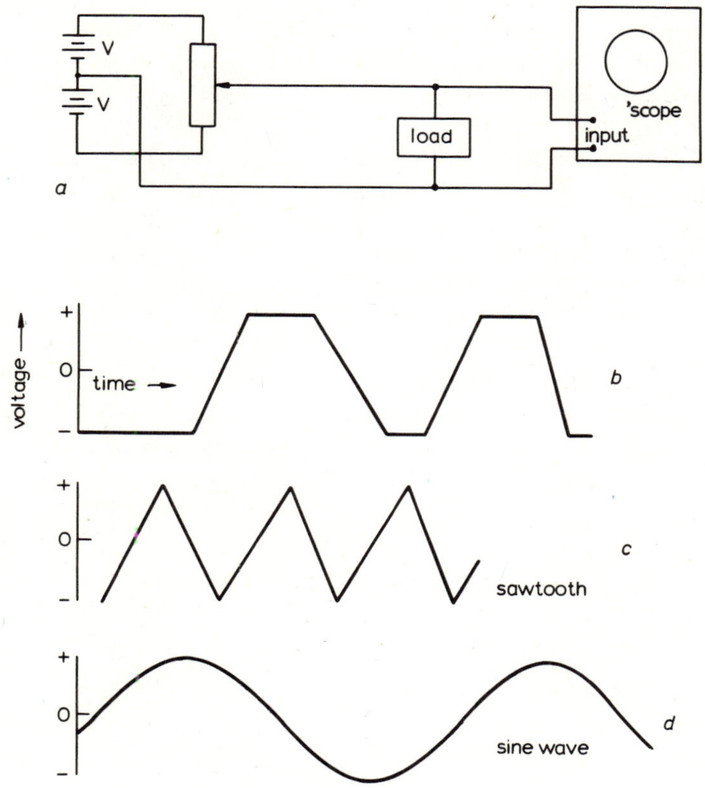

Fig. 3.2 Reversing d.c. supply to give various a.c. waveforms
a Oscilloscope circuit for producing waveforms
b Saw-square wave
c Sawtooth wave
d Sine wave

3.3 WAVEFORM DEFINITIONS

Many of the names and definitions used are interrelated but care must be taken to use the correct definition. A CYCLE is one complete train of events, as from A to E or B to F in Fig. 3.3, or even from X to Y if a particular intermediate voltage level is chosen. It is seen that there are many similar points between which the cycle may be defined, but the time or PERIOD of the cycle is the same in each example for the same waveform.

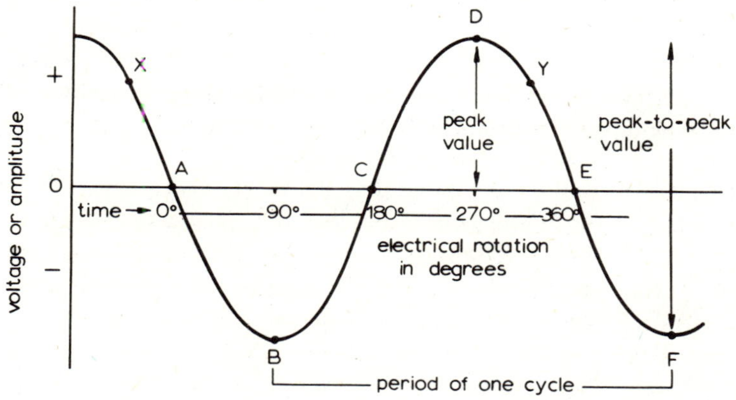

Fig. 3.3 The sine wave and some definitions

The time needed to complete one cycle, the period, allows the number of cycles-per-second to be easily calculated. The usual term for this is the FREQUENCY and if the period is, say 10 ms/cycle, i.e. (10/1 000) s/cycle, the frequency is 1 000/10, i.e. 100 cycles/s. The unit of frequency used to be cycles/s or c/s but has been changed to HERTZ or Hz, so that a cycle lasting 10 ms is seen to be part of a 100 Hz waveform.

So far the timing has been considered; the vertical axis of voltage (usually) has a deflection above and below zero. The zero is not seen since it is not traced out on the 'scope tube unless a double-beam type is available and one beam is left undeflected. The vertical size or AMPLITUDE of the trace may be referred to in two ways. Firstly, the PEAK value of the waveform, which is from zero up to D or down to B, where the positive peak value is equal to the negative peak value. And secondly, the PEAK-TO-PEAK value which is the vertical distance between B and D and is seen to be twice the peak value.

All the waveforms, Figs. 3.1 and 3.2*c* and *d* are of symmetrical waveforms where the positive side is a mirror image of the negative side, although shifted along one half cycle. A nonsymmetrical waveform such as Fig. 3.4 has a zero axis that is not central, and so the positive peak value is not the same amplitude as the negative peak value. The sum of the positive and negative peak values is, however, equal to the peak-to-peak value of the nonsymmetrical waveform.

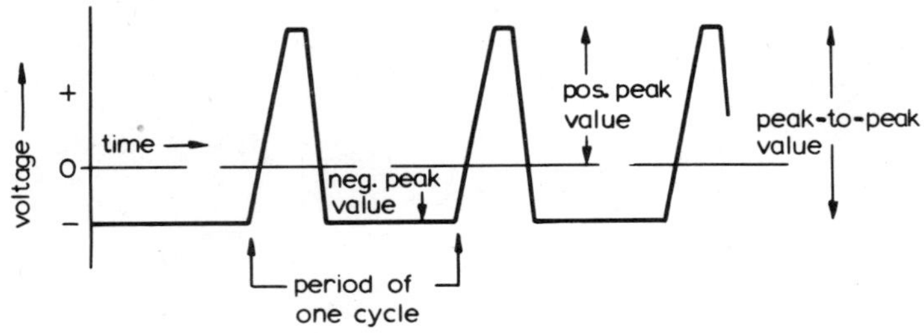

Fig. 3.4 Nonsymmetrical or asymmetric a.c. waveform

Any waveform that is not a pure sine wave is called a COMPLEX waveform, although special names may be given to pulse, square, triangular or ramp shape if there is a suitable title. The same terms of period, frequency, cycle and peak amplitude still apply to these complex waveforms.

3.4 PHASE

If a.c. is considered as being generated by a coil moving within a two-pole magnetic field, one complete cycle is generated for each complete revolution, i.e. 360° rotation. This means that particular points on any regular waveform – not necessarily a pure sine wave – may be referred to as the number of degrees after the start, 0° at a particular place on the waveform. In Fig. 3.3 the reference zero is taken as the waveform crosses the zero voltage axis, going from positive to negative. Thus the next time that the zero voltage axis is crossed by the waveform – in the other direction – is 180° away from 0°, and at 360° the waveform starts all over again.

When comparing two waveforms of the same frequency it is often found that they are not exactly in step or IN PHASE. Thus in Fig. 3.5. there are four waveforms, all the same shape but all reaching their maximum positive values at different times. The easiest way of referring to the time relationship of these waveforms is by their phasing; thus (*b*) is 90° (lagging) out of phase with (*a*); (*c*) is 180° out of phase with (*a*), also called phase reversed or inverted; while (*d*) is 270° (lagging) or 90° (leading) out of phase with (*a*).

Often in electronics there is a need to consider two signals when they are either in phase or completely out of phase, since these produce the conditions needed for special feedback systems.

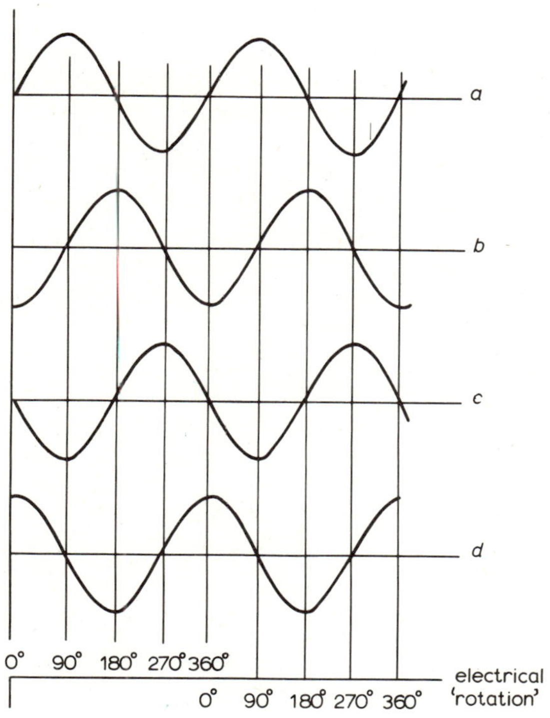

Fig. 3.5 Sine waves progressively shifted by 90° or a quarter cycle
a Original wave
b 90° (lagging) out of phase with *a*
c 180° out of phase with *a*
d 270° (lagging) out of phase with *a*

3.5 ROOT MEAN SQUARE OR R.M.S. VALUES

The mains electricity supply voltage is sine wave shaped, although an oscilloscope may show slight distortion in some localities owing to the loading effects of the apparatus being supplied. A very noticeable error will be found between the voltage reading of an a.c. voltmeter connected across the secondary of a step-down (for your safety) transformer and the scope waveform from across the same winding at the same time. The waveform shows a peak-to-peak voltage more than twice that of the voltmeter reading; even halving the waveform voltage to its peak value still produces considerable difference when compared with the voltmeter.

A large amount of electrical power is used for heating, and it may be necessary to change a soldering iron or lamp from an a.c. to a d.c. supply, depending upon which is available, and to still get the same *heating* effect in the load.

Now a d.c. supply is switched on and remains at full voltage until it is switched off, but an a.c. supply is continually changing, going through zero 100 times per second (for 50 Hz supplies). A special factor is therefore introduced whereby an alternating voltage that gives the same heating effect as a direct voltage is said to have the same value, the a.c. voltmeter being *calibrated* to read the same numerical value as the d.c. voltmeter. In reality, the peak value of the sine wave supply is 1·414 times as large as the d.c. supply. Hence the difference between the scope and the meter readings.

The 'equivalent heating' value is termed the root mean square or r.m.s. value and for a pure sine wave

1 V r.m.s. means 1·414 V peak or 2·828 V peak-to-peak
14·14 V r.m.s. means 20 V peak or 40 V peak-to-peak
(note that 1·414 is also $\sqrt{2}$, and you will appreciate the associated maths later on).

If the alternating waveform is not a pure sine wave, the conversion of peak to r.m.s. values is less easily done; it is worth remembering that virtually all a.c. voltmeters are calibrated to read r.m.s. values even though the dial or scale is not labelled to remind you of this.

3.6 A.C. LOADS

The complete electrical circuit includes a source of supply, two conductors and a load; a lamp, soldering iron or resistor are all transducers that convert electrical energy to heat energy and are affected similarly by both a.c. and d.c. sources, producing heat from both.

Apparatus involving electromagnetic effects however is often not operated in exactly the same way by d.c. and a.c. A d.c. type relay probably chatters or hums if a.c., even of the same r.m.s. voltage is applied, while a permanent-magnet motor or moving-coil instrument is sensitive to direction of current flow as well as amplitude – both of which are constantly changing with alternating supplies – and so the motor or instrument pointers just hum or dither about zero and produce no useful movement or indication.

A change of design produces suitable a.c. transducers to be used as loads, but for simple signalling along wires, a.c. is little different from d.c. if only on-off codes are considered. With the coming of electronics however (with amplifiers, different systems and transducers) a.c. becomes of far greater use than d.c. and so will be considered in greater detail.

Chapter 4

Alternating waveform generation and transmission

4.1 SOUND

As human beings, we are extremely complicated systems and have at least five senses with which to either receive or transmit communications. Some of the sensing systems, taste and smell, operate in a way that is still not properly understood, but feelings, sound and sight have been quite fully investigated for signalling purposes.

We make a sound with the mouth, tongue and throat acting together to vary the flow of air from the lungs; try speaking without allowing any air at all through the throat. Or try speaking without moving the lips, tongue or throat but allow the air out this time. The mechanical movements without airflow or vice versa produce few recognisable sounds!

This process of adding information – for that is what any word or sound is – is called MODULATION, and our speech consists of variations in (or modulation of) air pressure above and below the average air pressure existing in the locality. Flapping a stiff piece of card in the air two or three times a second will produce no sound as far as the hearing is concerned because the frequency of air-pressure variations are too low. As the frequency is increased we start to 'hear' the sound (as well as feel the draught) at about 15 flaps per second or 15 Hz. Few people can talk at such a low frequency – it is more of a growl or grunt – but we hear 'sound' from about 15 Hz up to about 15 kHz.

Our sound receivers, the ears, convert the modulated air pressure into mechanical movement of small bones, skin and fibres within the ear; nerves are then triggered into operation, electrical impulses travel to the brain along the nerves and at the brain the received information is 'decoded' and passed on to the appropriate part of the body for further action.

The hearing thus depends upon having ears that are not blocked up and within the ears the fibres and nerves must be able to respond to small variations of air pressure. As we get older the whole hearing system hardens and this accounts for increasing deafness (insensitivity to small sound) and also the loss of sensitivity to high-frequency notes and complaints of 'muffled' speech and music.

If you whistle through your lips very gently, the sound is 'pure' and the variations of air pressure would be almost a pure sine wave. As you blow harder – still on the same note – the sound becomes more shrill and piercing. The same basic frequency is still being produced (the same note) but the waveform has been altered to give a different characteristic or 'tone' or 'timbre'. It is these differences from the pure sine wave that enable you to recognise the difference between unseen speakers' voices that might be saying exactly the same words, even at the same frequency, yet still belong to different persons.

4.2 TRANSDUCERS (Fig. 4.1)

Basically there are only two transducers to think of when considering sound. If sound into the transducer produces an electrical output the transducer is called a microphone. If an electrical input produces sound then the transducer is called a loudspeaker. Variations exist of course; the record or disc pickup is not quite the same as a microphone while an earphone or earpiece is not quite the same as a loudspeaker.

At the microphone the variations in air pressure move a diaphragm in sympathy with the pressure waveform; the diaphragm movements generate voltage variations of the same waveform. Thus the waveform of the sound is reproduced as the same waveform voltage but this does NOT mean that you can speak or whistle a voltage!

At the loudspeaker, current and voltage variations cause vibrations of the diaphragm or cone to produce the same waveform air pressure variations – but as above, you cannot hear a voltage or current!

We cannot see the sound directly – even with a coloured gas that becomes thicker and thinner, because our eyes are so slow to respond that we would never realise that the colour density was varying – but we can speak into a microphone (or a loudspeaker acting 'backwards') and see the resulting voltage variations displayed on a scope. The louder the sound, the larger the amplitude of the voltage waveform; the higher the pitch or note, the greater the number of cycles displayed in a given time. And finally, the shape, tone or timbre of a note varies the actual

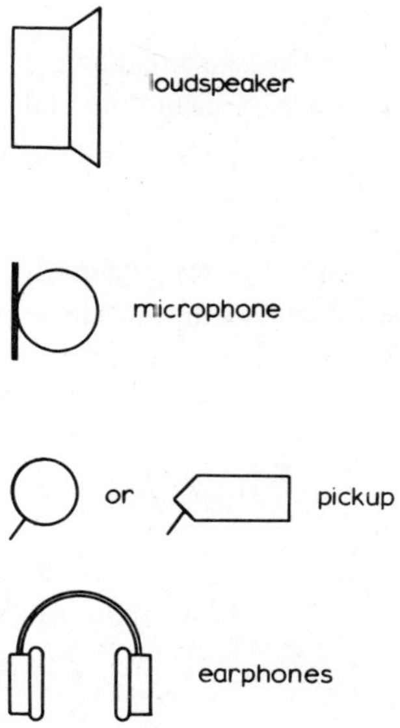

Fig. 4.1 Transducer symbols

waveform of any one cycle from a pure sine wave through countless individual variations.

Although we say we can hear from 15 Hz to 15 kHz, we normally use a much more limited range of from 150 Hz to about 5 or 6 kHz for speaking and singing and an even more restricted range of from 300 to 3 000 Hz for reasonably intelligible speech.

A frequency spectrum is shown in Fig. 4.2 for a range of instruments and voices.

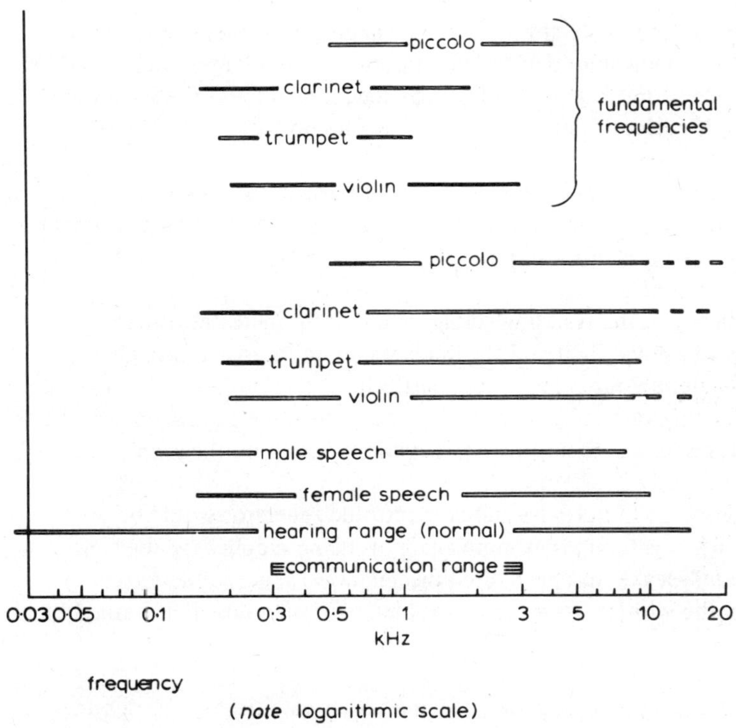

Fig. 4.2 Range of frequency used by musical instruments and voices

4.3 ELECTROMAGNETIC SPECTRUM

About a century ago it was found that current flowing in a conductor produced a magnetic field, also that a variation of magnetic field about a conductor generated a voltage in the conductor. Thus variations of magnetism and electricity exist together and variations in the waveform of one (larger or smaller than average) affect the waveform of the other.

Consider a simple bicycle generator (often incorrectly called a dynamo), where the road wheel rotates a cylindrical magnet within a coil, and also consider that the output of the generator is to be connected to lamps or loudspeakers or just wires as required. Fig. 4.3 gives the circuit, showing how the various loads could be switched on.

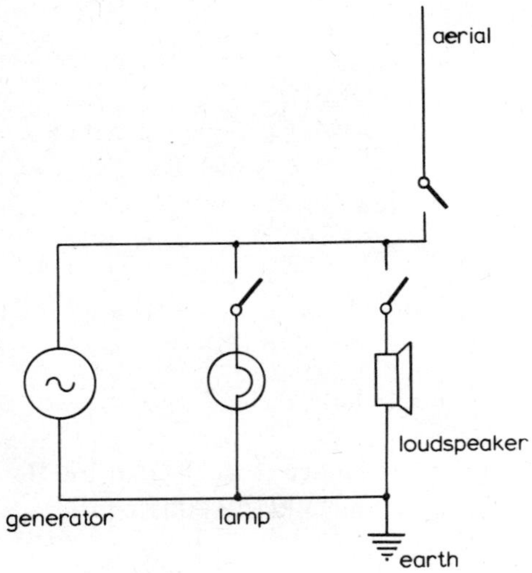

Fig. 4.3 Variable frequency feeding different loads

As the magnet is rotated by the wheel – slowly at first – the alternating voltage output would make a lamp bright-dark-bright-dark and so on, each brightness being due to a single half cycle of the voltage waveform at low frequencies. A loudspeaker cone could be seen to be moving outwards-centre-inwards-centre at the same time, showing the difference in the polarity of the half cycles but otherwise producing no noise.

At about 15 revolutions per second, or 15 Hz, the lamp appears almost uniformly bright because of the eye's slow response to variations in light output and also the heat storage of the filament. The loudspeaker cone appears to be stationary now but starts to produce a very low-pitch musical note.

As the rotation frequency and hence the frequency of the resulting voltage and current increases through 100 Hz – 1 kHz – 10 kHz – 15 kHz, the lamp still stays alight and the loudspeaker emits an increasingly high pitched musical note or sound until (depending upon your age and your ears!) at 20 kHz the 'sound' is no longer there but the lamp is still alight. The air is still being vibrated as may be seen by connecting a nearby microphone to an oscilloscope, when a voltage waveform transmitted via air-pressure variations is seen on the scope screen.

It is unrealistic to use a simple generator at these high frequencies and an electronic generator would be used instead. As the frequency is increased still further it is found that more energy is being supplied by the generator than is accounted for by the lamp alone (the loudspeaker has been disconnected now) and a suitable receiver would show that energy is being radiated from the wiring into space via magnetic and electrical field variations. 'Radio' frequency radiation is occurring.

Our bodies do not detect these radiated radio frequencies which may become higher and higher – until our skin and nerves detect a range of frequencies as heat; then the eyes detect a red glow, changing as the frequency increases to orange, yellow, green, blue, indigo and violet (colours of the rainbow), then to ultraviolet, detected by a change in skin colour (sunburn), and then progressing even further up towards X rays and so-called atomic radiation where the emissions are able to go through 'solid' objects.

Frequency range	Name	Uses
0 Hz	Direct current; d.c. zero frequency; z.f.	
20 Hz – 20 kHz	Audio frequency; a.f.	Speech, music, hearing
* 300 – 3000 Hz	Communication channel for speech 'information'.	
30 kHz – 300 kHz	Low frequency l.f. Long wave l.w.	Domestic/Entertainment broadcasting
300 kHz – 3000 kHz 3 MHz	Medium frequency m.f. Medium wave m.w.	
3 MHz – 30 MHz	High frequency h.f. Short wave s.w.	Long-distance broadcasts and radio transmission
30 MHz – 300 MHz	Very high frequency v.h.f.	Television broadcasts, mobile radio, f.m. broadcasts
300 MHz – 3000 MHz 3 GHz	Ultra high frequency u.h.f.	Television broadcasts, mobile radio, aircraft beacons, radar
3 GHz – 30 GHz	Super high frequency s.h.f.	Radio links, radar
⋮	Infrared	
	Light: Red, Orange, Yellow, Green, Blue, Indigo, Violet	Seeing
⋮	Ultraviolet	
	X-rays 'Atomic' radiation	Medical, engineering diagnosis

The useful range of frequency shown in Fig. 4.4. is thus enormous – it is emphasised again that you do not sense a voltage variation – increasing up to the gigahertz (thousands of megahertz) region for radar use before increasing even further up to 'light' frequencies. Luckily a lamp bulb is a convenient and cheap transducer of zero frequency (d.c.) and lowish-frequency a.c. supplies to light and heat, and our eyes are conveniently portable receivers for getting the message.

4.4 COMPLEX WAVEFORMS

As already stated, if a waveform is a pure sine wave then it consists of one frequency only; if a waveform is complex however it may be proved to consist of the basic frequency or FUNDAMENTAL frequency sine wave plus assorted HARMONIC frequencies. A harmonic is two, three, four or more times the basic frequency all the way up to infinity times the fundamental. Note that a harmonic must be a whole number (or integer) times the fundamental – you cannot specify a harmonic as 'two and three-sevenths' times!

In Fig. 4.5 is shown a fundamental frequency, and, of rather smaller amplitude, the second harmonic (i.e. twice the frequency of the fundamental). By considering all the parts of the waveforms above the zero line as positive, below the zero being negative, suitable addition of positive values or sometimes positive and negative values at any single instant will give resulting INSTANTANEOUS values for the RESULTANT waveform.

The points may be joined to show the result of the addition, and it is seen that the addition of fundamental plus second harmonic frequencies produces a complex waveform with the same basic frequency but a new shape.

To the musician it is quite common to deliberately play several notes together to produce chords, some pleasing or some discordant and unpleasing. The characteristic difference between, say, a flute and a bugle playing the

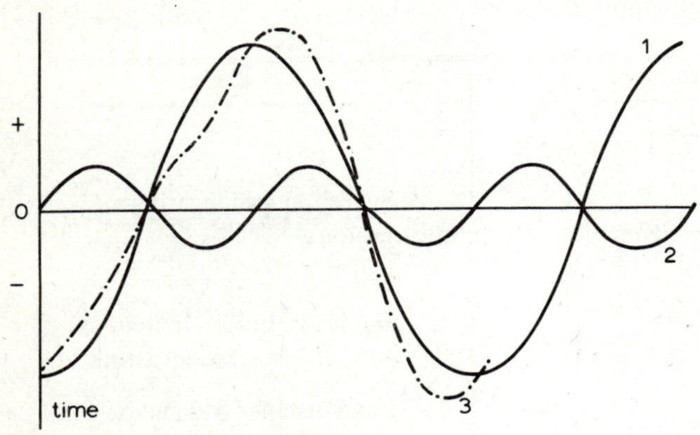

Fig. 4.5 Fundamental and second harmonic added to form complex waveform
1 Fundamental sine wave
2 Second harmonic of 1
3 Addition of 1 and 2 to form complex waveform

the same note is due to the automatic production of harmonics which make a flute sound like a flute and not like a bugle.

4.5 THE SQUARE WAVE

One of the special waveforms that occurs over and over again is the pulse, square or rectangular waveform, shown in Fig. 4.6*a*. The fundamental frequency of Fig. 4.6*b* is the same as the pulse frequency, but for sufficient extra

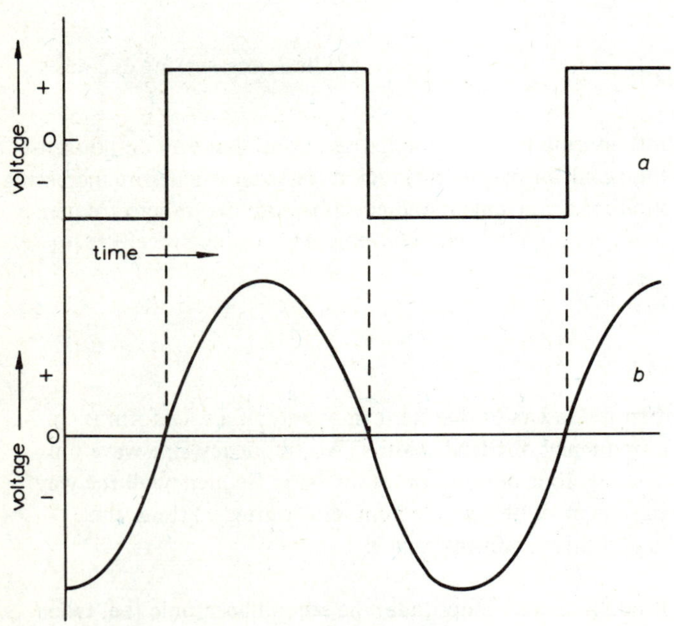

Fig. 4.6 Sine and square waves of same frequency

harmonics to be added to square out the corners it is found that there need to be added all the ODD HARMONICS TO INFINITY – in the correct proportions of course, (remember this). For instance a 200 Hz square wave would consist of

	200 Hz sine wave	(fundamental)
plus	600 Hz sine wave	(3 x fundamental)
plus	1 000 Hz sine wave	(5 x fundamental)
plus	1 400 Hz sine wave	(7 x fundamental)
plus	1 800 Hz sine wave	(9 x fundamental)
plus	2 200 Hz sine wave	(11 x fundamental)
plus	12 600 Hz sine wave	(63 x fundamental)
plus		

If the correct proportions and correct frequencies are not present, then the square wave stops being square and probably starts to have rounded corners or some other distortion away from the original shape. Luckily, although an infinite frequency spectrum is mathematically required for perfection, the first ten or so harmonics will have a reasonably square shape. This range of frequencies would mean light frequencies and beyond, and they do not travel via solid conductors too happily!

Square waves are sometimes used as test inputs for amplifiers to see if the amplifier can handle the very large range of frequencies involved all at the same time; any resulting distortion at the output tells the designer or tester about the goodness of the amplifier and its suitability for 'hi-fi' or otherwise.

4.6 VELOCITY OF WAVES

When one throws a ball or fires a bullet, time is required for the ball or bullet to travel the distance. A sound also requires time for a single air compression to move away from the source of sound, followed by a rarefaction, followed by succeeding pressure and rarefaction waves as the sound is produced. The velocity or speed of the sound away from the source depends upon air temperature and pressure mostly and is about 330 m/s.

Electromagnetic waves however travel much faster, at 300 000 000 m/s; so fast that in a millionth of a second light travels 300 m. Yet nowadays we can measure such small time intervals and use such high frequencies that even a millionth of a second is a long time period if ultra high frequencies (u.h.f.) – 625 line television programmes – are considered. At around 600 000 000 Hz or 600 MHz, 600 individual cycles are formed and transmitted in a millionth of a second. In the 300 m travelled by the start of the first cycle's electromagnetic field there will be strung out behind that first portion through 300 m of space the other 600 cycles, all travelling away from the source at the same speed.

With a little imagination you can see that each cycle will occupy half a metre and this introduces the relationship between frequency and wavelength, where frequency (in Hertz) x wavelength (in metres) = 3×10^8 m/s.

A radio transmission of 200 kHz may also be marked on the tuning dial as 1 500 m since 200 00 x 1 500 = 3×10^8; a transmission tuned in at 300 m on the dial is similarly calculated to be of 1 MHz frequency since 1 000 000 x 300 = 3×10^8.

4.7 REVISION

Hearing: average range is from 15 Hz to 15 000 Hz or 15 kHz for humans; bats and dogs can hear higher frequencies than we can, hence the use of the 'silent' dog whistle.

Speech: men use about 80 Hz to 8 kHz, women's higher pitched voices are probably using about 150 Hz to 10 kHz.

Singing: up to double the high speech frequencies.

Music: covers the whole range of hearing frequencies.

Communication, i.e. telephone quality of understandable speech, needs 300 – 3 000 Hz.

Sound: air pressure variations at the listener's ear.

Audio frequency voltages and currents are not sound until changed by a suitable transducer.

Radio frequency energy may be radiated through space without needing conductors or air.

Sound waves travel with a velocity of approximately 330 m/s through air while electromagnetic waves travel through space with a velocity of 3×10^8 m/s.

An instantaneous set of measurements of maximum and minimum energy levels taken of radiated energy moving away from a source will show 'wavelengths' to exist; wavelength and frequency are related by:

$\lambda \times f = c$ where λ is wavelength in metres
f is frequency in hertz
c is velocity of electromagnetic propagation, 3×10^8 m/s.

Chapter 5

Alternating waveform reception

5.1 THE SENSES

The use of man's senses has already been mentioned; the ears respond to a range of frequencies of air pressure variations while the eyes respond to a very narrow band of electromagnetic frequencies that are called 'visible light'. The nerves respond to the low frequency end of the visible light range, the red and infrared rays, by feeling warm; at the high frequency end the skin reacts to form a protective red then brown barrier to ultraviolet rays. The eyes must be protected from ultraviolet rays, otherwise extreme discomfort and then blindness occurs. This is the reason for wearing goggles while arc welding where much ultraviolet radiation is produced.

5.2 SIMPLE RADIO TRANSMISSION

At up to 15 to 20 kHz there is very little radiation of electromagnetic energy from conductors or 'aerials' unless the aerial is extremely long, perhaps several thousand metres. Higher than this frequency the energy is radiated more and more easily and reception at a distance involves the erection of an aerial to intercept some of the energy and then 'detect' the presence of the energy with a receiver. An early experimental 'wireless' transmission was performed by Hertz whereby a spark was made to jump across a gap between the rods of the transmitter's dipole aerial as in Fig. 5.1. A similar receiving dipole aerial nearby had induced sparks jumping across a far smaller gap –

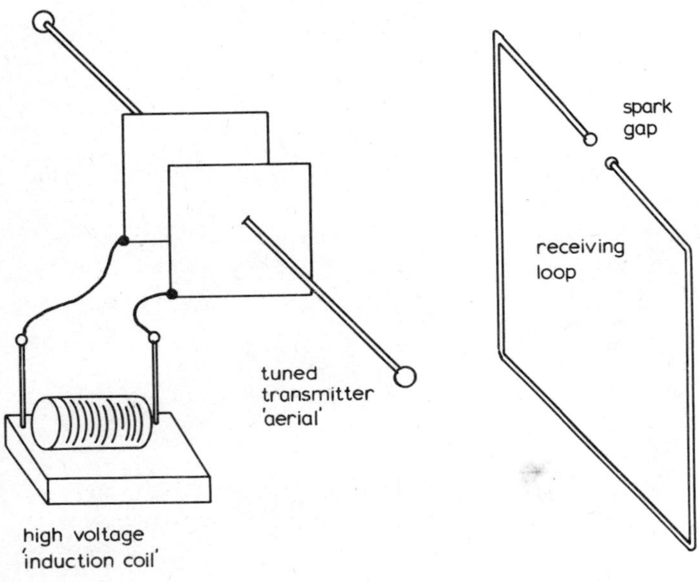

Fig. 5.1 Transmission of 'wireless' waves by Hertz

so in this case the final detector was the eye, seeing the received spark. At greater distances sensitive voltage detectors used in conjunction with rectifier or 'detector' diodes could be used. Even greater ranges are possible if an amplifier is used to drive the detecting device.

5.3 TRANSMISSION RANGES

The range of sound depends mainly upon the loudness or amplitude of the source – a shout obviously carries further than a whisper because as the initial amplitude of pressures and rarefactions becomes smaller with distance from the source, the limit of hearing sensitivity determines when the 'sound' falls below a detectable level. For a really large sound, consider that when the top of the volcano Krakatao blew itself apart in 1903 with an enormous blast, the sound was reported to have travelled around the world at least twice; an atom bomb or thunder storm makes a comparatively small bang!

Thus for long range communication, very high power sound sources would be needed, and the nearby listeners deafened. Radio waves travel much more easily through air or space, are unobtrusive to people near the source, and, with the aid of electronics, are easily detected at the receiver position, even though that position is many thousand metres distant.

Light communication range has always been taken as straight line or line-of-sight but the development of lightguides made with glass fibres has led to considerable rethinking since the lightguide may now be treated very similarly to cables when turning corners and travelling underground. The receivers are of course all electronic and complicated.

5.4 NEED FOR CARRIER

If a pair of socks or a t.v. receiver is manufactured, unless the customer travels to the factory and takes away his purchase, the items must be conveyed to the final customer by train, lorry, or messenger – a CARRIER. The customer, having accepted the socks or t.v. receiver no longer needs the carrier which is then dismissed.

Information such as speech or music that is to be transmitted (delivered) to a distant receiver must be 'loaded' onto a carrier. Even when the sound information is transduced into information voltage the frequencies in use are too low to achieve long distance radiation by radio methods, so a radio frequency CARRIER is used and the information then becomes the MODULATION of the carrier. At the receiver the modulated carrier is DEMODULATED and the original information regained.

5.5 FILTERS

Any receiver must incorporate some form of filter to select the wanted information and reject other information offered but not wanted. The hearing, in conjunction with the brain, can filter or select one particular person's voice from three or four speakers. The eyes, again using the brain, can select one particular colour (frequency of light emission or reflection) from many colours. A t.v. receiver is 'tuned' to select one bunch of information – audio plus video, a programme – from the several lots of information of various frequencies offered on the different channels.

All the systems mentioned are using filters to get rid of distracting information, or in other words to select the required frequencies from an extremely complex waveform offered.

'Tone controls' on audio amplifiers operate in a similar fashion except that here the liking of an individual person for 'bright' or 'boomy' music allows either the low or high frequencies, respectively, to be filtered away (and lost) to leave the wanted characteristics of sound.

A final test that you can do with your own voice – cheaper than buying a tape recorder – is to place two magazines with the edges against your cheeks in front of your ears. This effectively blocks the direct and nearly instantaneous path of your own voice to your own ears. Now speak, and you hear the voice that other people hear. Not so loud because of the greater distance travelled by the sound waves. From walls and ceiling there are echoes that are normally masked by your louder direct voice. A different voice – but it is you – that is often blamed on the microphone or tape recorder when people first hear the voice to which others have to listen. The room and furnishings are acting as a form of filter.

Chapter 6

Modulation

6.1 MODULATION METHODS

A train of cycles viewed on a scope looks like Fig. 6.1 *a*. If the setting of the time base, the rate of sweep across the tube in the horizontal or X direction, is low the individual cycles are so close together and cramped up that the display looks like a horizontal blur as in Fig. 6.1 *b*. This blur is enclosed in the ENVELOPE although as you will see later the envelope need not have a horizontal top and bottom to it.

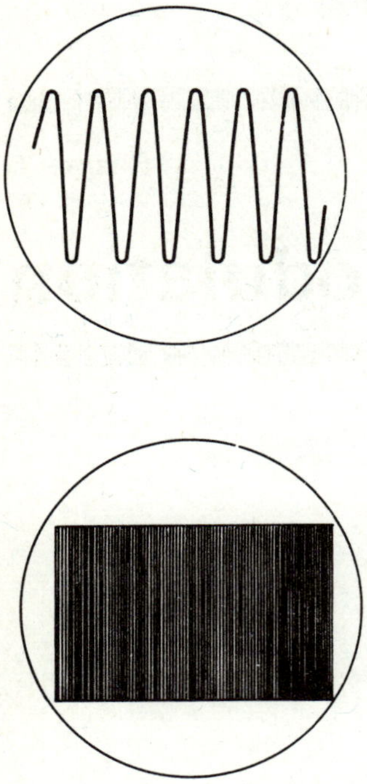

Fig. 6.1 R.F. waveform
a producing an envelope
b when viewed with slower running timebase

The carrier used to assist the passage of information from the sender to the receiver will be a radio frequency (r.f.). A scope can view radio frequencies as easily as audio frequencies (a.f.) and only by noting the time-base speed can the difference in frequency of a train of steady sine waves be ascertained.

Now information is to be modulated onto this train of r.f. cycles and there is a limited number of ways available:

(i) The r.f. may be stopped and started – probably using the Morse code of long and short dashes and dots – so that the receiver interprets the code back into letters, words and the original information. This is CONTINUOUS WAVE or c.w.
(ii) The size or AMPLITUDE of the r.f. may be varied above and below average by a.f. information and at the receiver the AMPLITUDE is continuously monitored and then demodulated into the a.f. information. This is AMPLITUDE MODULATION or a.m.
(iii) The FREQUENCY of the r.f. may be varied above and below the average or centre frequency by the information and at the receiver the FREQUENCY is continuously monitored and demodulated into the information. This is FREQUENCY MODULATION or f.m.
(iv) The PHASE of the individual r.f. cycles may be shifted by the information backwards and forwards in time relative to the phase of a reference oscillator. The result is very similar to the f.m. process and will not be dealt with further. This is PHASE MODULATION.

6.2 CONTINUOUS WAVE OR C.W.

The title is confusing; it is obvious that to generate and transmit an unchanging and continuous r.f. signal will inform the receiver of nothing except that the sender has switched on! The title, therefore, refers to the fact that the method of modulation is NOT f.m. or the normal type of a.m.

C.W. was the first method to be used and was as simple as shown in Fig. 6.2*a*. The r.f. generator was at first just an oversized electric bell mechanism with a secondary winding, that had a high voltage induced into it, and then connected to the radiating aerial. The key or on-off switch could either switch the bell make-and-break mechanism or could be connected as shown between the generator and the aerial; Fig. 6.2*b* shows the envelope of a short c.w. message.

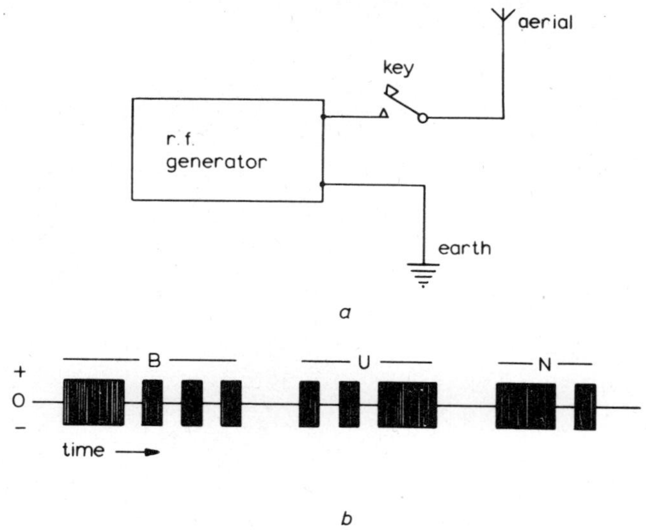

Fig. 6.2 (*a*) C.W. transmitter
(*b*) waveforms at aerial

This crude method was used up to about 1920 and produced a very raspy note in the receiver earphones; both transmitter and receiver were very bulky, and the range was fairly short. Nowadays the r.f. generator would probably be a transistorised oscillator circuit – and the receiver would contain many transistors and be much more sensitive than was possible 50 years ago. C.W. is still in use as a method; it is simple and reliable but rather slow and requires good operators to code and decode the information at each end of the information chain.

6.3 AMPLITUDE MODULATION OR A.M.

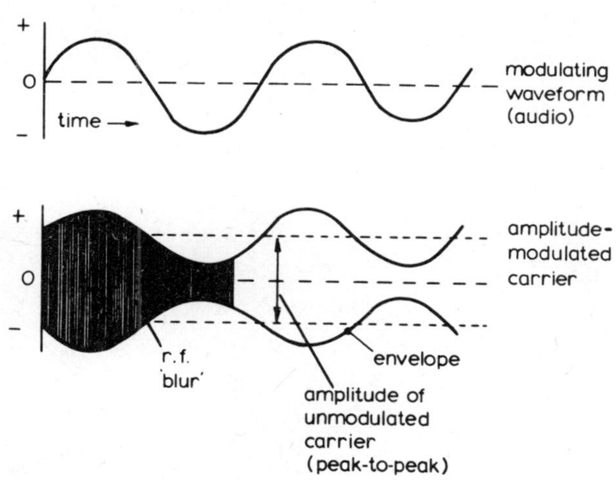

Fig. 6.3 Audio waveform and resulting amplitude-modulated carrier envelope

A more complicated system than c.w., the carrier is present the whole time but is made to vary in amplitude in sympathy with the modulating waveform shown in Fig. 6.3; in this instance an a.f. sine wave is used for modulation.

The r.f. is shown as a blur because it is so much higher in frequency than the modulating a.f. that the individual cycles of r.f. merge one with the other and only the tips at top and bottom trace out the envelope of a.f. The transmitter block diagram of Fig. 6.4 once again has an r.f. generator stage as was used for the c.w. transmitter but this time the output has to be varied in amplitude instead of switched on and off. The microphone converts air-pressure variations – sound – into small voltage variations of the same waveform. These small voltages are made larger by the a.f. amplifier and then fed to the stage wherein the r.f. amplitude is varied by the a.f.

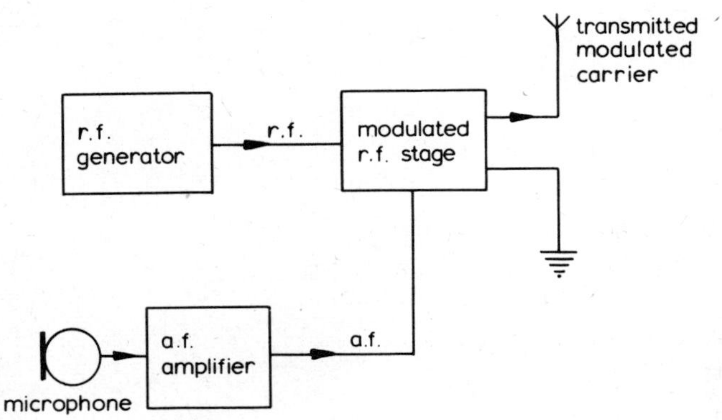

Fig. 6.4 Block diagram of amplitude-modulated transmitter

The modulating stage could be likened to a variable resistor controlled by the a.f., allowing more or less than average of the r.f. to travel through. The modulated r.f. is then fed to the aerial and earth system.

Fig. 6.3 shows the relationship between an audio waveform and the corresponding modulated r.f. Note that since a cycle of r.f. voltage varies above then below zero, as the overall amplitude of the r.f. is made larger and then smaller the a.f. curve traced out by the positive tips of r.f. is a mirror image of the curve traced by the negative r.f. tips. If the modulating audio is not a pure sine wave but a typical complex waveform, then the envelope of the carrier displays this complex waveform on both sides. Note also that the carrier must never be allowed to disappear altogether since for that period of time there would be zero information radiated.

6.4 SIDEBANDS

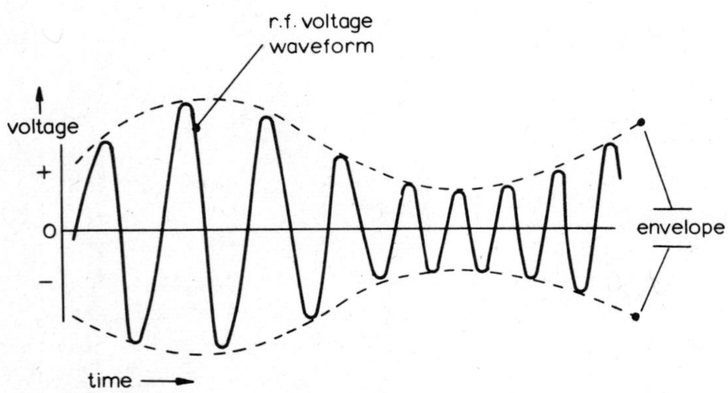

Fig. 6.5 Expanded a.m. waveform

In Fig. 6.5 is shown an expanded view of a modulated carrier. The waveform is not a pure sine wave and it has already been stated that an impure or complex wave consists of the fundamental plus other sine waves. In this instance it appears arguable whether the low audio frequency is the fundamental; the frequency components of the modulated carrier are found to lie both below and above the carrier frequency as described.

Consider a modulating audio frequency of 500 Hz and a carrier frequency of 300 kHz or 300 000 Hz. After modulation the frequencies present in the complex waveform are found to be:

299 500 Hz i.e. carrier minus modulation frequency
300 000 Hz i.e. carrier alone
300 500 Hz i.e. carrier plus modulation frequency

And this reasoning applies to all a.m. systems; algebraically if f_a is the audio frequency and f_c is the carrier frequency, then the frequencies present at the output of the modulating stage are

$f_c - f_a$ i.e. the LOWER SIDE FREQUENCY
f_c the CARRIER
$f_c + f_a$ the UPPER SIDE FREQUENCY

Note that there are no audio frequencies transmitted, all the frequencies are r.f.

The frequencies are often displayed as you would identify them one at a time with a superselective receiver along the tuning dial as in Fig. 6.6, a frequency spectrum of the radiated r.f.

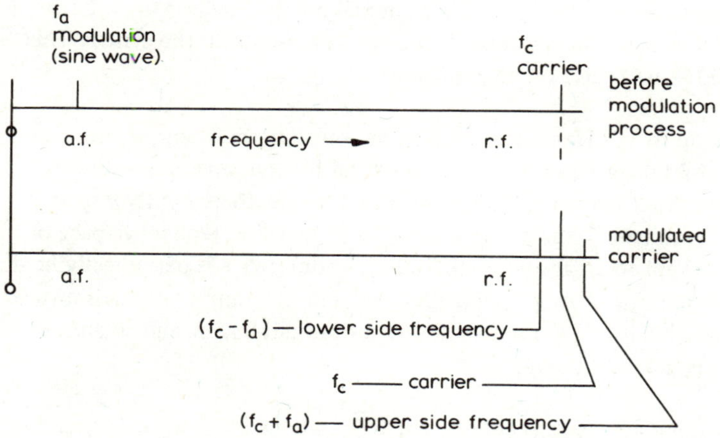

Fig. 6.6 Frequency spectrum of a.m. before and after modulation showing side-frequency locations

Now f_a will rarely be a perfect sine wave but in the instance of speech or music will be a complex waveform in itself. This means that added to and subtracted from the carrier frequency there will be a number of frequency components below and above the carrier rather than just single side frequencies. The band of frequencies making up f_a therefore results in the production of upper and lower SIDEBANDS on either side of the carrier – the carrier still remains as a single frequency. Fig. 6.7 shows the 'superselective' receiver scale or spectrum and the places where signals are heard – the receiver would be so selective that it would detect only one frequency at a time and would not reproduce the speech or music.

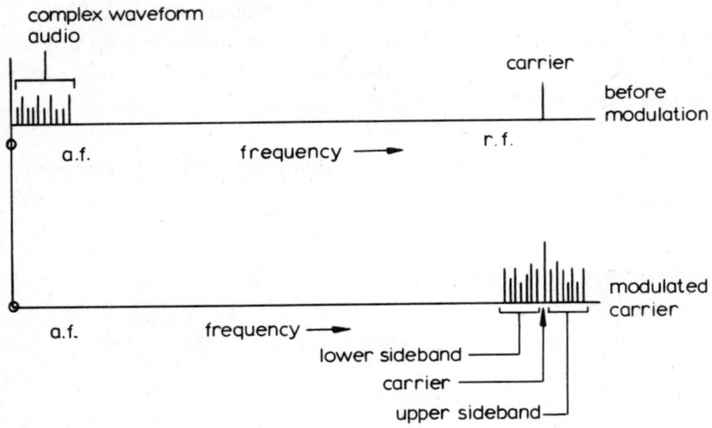

Fig. 6.7 Frequency spectrum of a.m. system; complex audio-modulation producing sidebands

6.5 BANDWIDTH

The a.m. transmission carrying the audio in the form of upper and low side frequencies or sidebands ought to be received complete if the modulation is to be recovered at the receiver. This means that a superselective receiver is not really suitable but that a receiver able to receive a band of frequencies is required. The BANDWIDTH of the transmission is from the bottom of the lower sideband to the top of the upper sideband, i.e. twice f_a and this is the bandwidth that must be accepted by the receiver.

The importance of this bandwidth consideration lies partly in the number of stations or channels that may be placed side by side in any one waveband. The medium waveband used for entertainment on most receivers lies between 500 and 1500 kHz. Although this is less than the 'official' medium wave (m.w.) band of 300 to 3000 kHz; the other frequencies are reserved for special purposes and there is no point in making the receiver tune the whole of this band to listen to lots of squeals and whistles with the entertainment portion cramped in the centre. Thus the normal total bandwidth is from 500 to 1500 = 1000 kHz on medium wave.

Suppose that good quality speech and music of up to 10 kHz maximum is to be transmitted. Then the upper and lower sidebands produced by the a.m. process will take up 20 kHz bandwidth total for the programme and so fifty different stations could exist side-by-side without overlapping. But, in Britain alone, there are twenty or more stations on m.w. – and then there is the Continent – and a few 'pirates' – so that the frequency space or bandwidth occupied by any one station must be kept to an absolute minimum. A conference is periodically held in Copenhagen for countries to have their particular carrier frequencies allocated and the number of transmitters allowed per country is also decided. The bandwidth allowed per a.m. station is at present 9 kHz, and so about one hundred stations have room on medium wave for side-by-side operation.

Listen to medium wave any evening or night – there are stations crammed in everywhere, even on top of each other and it is obvious that there are many more stations than could be allowed at only one per 9 kHz channel. Luckily in the daytime the transmission range is relatively short on m.w. and one channel may be allocated to up to twenty low power stations that are geographically separate without undue interference, but the restricted bandwidth and the interference from adjacent or sharing channels is one of the reasons for transferring programmes to the v.h.f. bands where possible.

6.6 SINGLE SIDEBAND

Looking at the radio frequency spectrum of an a.m. transmission in Fig. 6.7 it is seen that the audio information occurs twice, once in the upper and again in the lower side frequency or sideband. If the transmitter is made a little more complicated and one sideband is stopped by a filter then the receiver will have a single sideband and the carrier received by the aerial and will still be able to demodulate these frequencies into speech or music.

With even more complication at the transmitter, the carrier may also be stopped by a filter, leaving only a single sideband to be transmitted and received. The receiver now becomes considerably more complicated as well and must reinsert a locally produced 'carrier' so that the demodulation process may be carried out to produce speech and music.

The advantage of a single sideband or s.s.b. transmission (with or without carrier) is that only half the bandwidth is needed. The disadvantage is the more complex circuitry needed, particularly at the receiver end and the very careful tuning needed for good reproduction of speech. S.S.B. is, however, now being used for maritime communications and it is probable that even m.w. will have to change to s.s.b. rather than the present full double sideband or d.s.b. if the need for more stations continues to increase.

Interference on a.m. transmissions has always been a nuisance, particularly the 'impulse' type of interference resulting from the use of electric razors, trains, trams, flashing Christmas lights and most types of on-off switching. The receiver for a.m. relies upon a change in amplitude of the station to which it is tuned for the demodulator to produce sound. Now a sudden on-off switching of either a.c. or d.c. to a load will be part of a square wave pulse, and the harmonics (all the odd ones) that go towards making this square wave will almost certainly possess one or two frequencies which lie within the bandwidth of the station to which the receiver is tuned. So the receiver reproduces a click if a lamp is switched on or off, or there is a buzz if an electric drill is being used locally – and as for trams or full-size electric railways, the interference pulses are enormous and their range is up to a mile or more.

There is little hope of eliminating pulse interference from a.m. transmission except by hoisting the aerial as far into the sky as possible to receive plenty of r.f. and then screening the down lead from local interference pickup.

6.7 FREQUENCY MODULATION OR F.M.

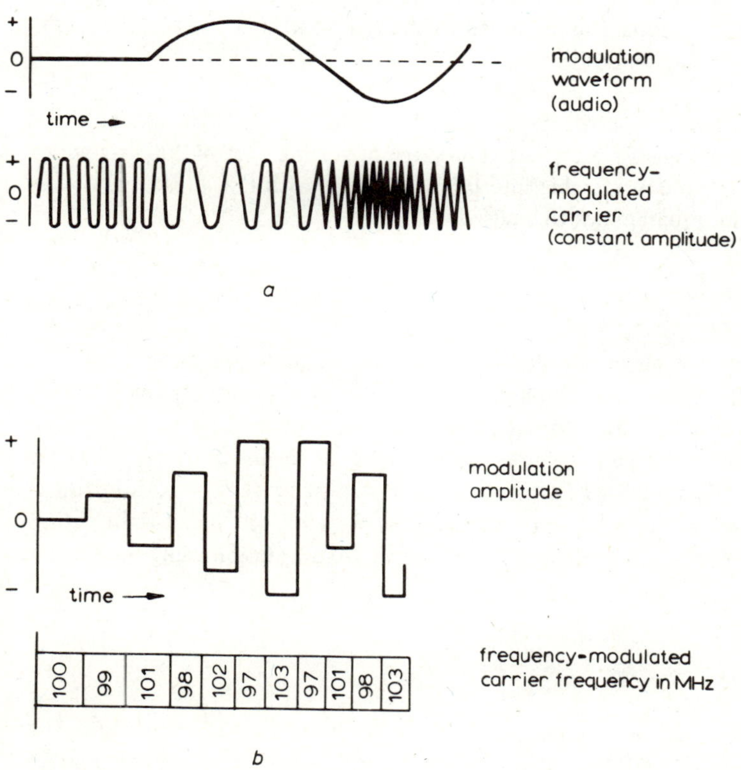

Fig. 6.8 Frequency modulation
a showing the variation of carrier waveform
b frequency deviation

An American, Major Armstrong of the US Navy, was one of the first to analyse and recommend the use of f.m. but the electronics then available, about 1930, were barely able to cope with the problems, particularly in the receivers.

The carrier frequency for the transmitter is generated by an oscillator, but this time the amplitude is maintained constant and the frequency is varied above and below the centre carrier frequency by the waveform of the audio frequency. This variation of carrier frequency is called CARRIER DEVIATION and the louder (bigger amplitude) the audio is, the greater the deviation of the resulting modulated carrier. The higher the frequency of the audio signal, the faster that the carrier deviates above and below its nominal frequency.

An oscilloscope display is quite difficult to produce, and so even is a realistic drawing. Fig. 6.8 *a* shows a sketch of a cycle of audio and the resulting frequency modulated carrier, but a better display is shown in Fig. 6.8 *b* where a square wave audio or modulation waveform is shown with the resulting frequencies of the f.m. transmission, rather than the waveforms of the carrier. The nominal or centre carrier frequency is taken as 100 MHz.

6.8 SIDEBANDS AND BANDWIDTH OF AN F.M. SYSTEM

Now a change from one frequency to another, particularly with square wave modulation or high frequency audio, demands some almost instantaneous changes of slope within the constant-amplitude carrier and as a result the bandwidth needed for f.m. is considerably greater than for a.m. Also, there is no simple relationship between bandwidth and modulating frequency, since the bandwidth also depends upon the carrier deviation, i.e. upon the loudness of the modulation.

In theory, the bandwidth is almost infinite, with sidebands and side frequencies stretching along the frequency spectrum on either side of the carrier to zero frequency and to infinity. In practice, the deviation (and hence the bandwidth) is set up by a technician working to the equipment manufacturer's specifications, and the bandwidth of mobile and point-to-point radio telephone type apparatus is limited to about 12·5 kHz at present, the voltage amplitude of the theoretically large sidebands being reduced to insignificance.

The bandwidth of the v.h.f. entertainment programme, however, is typically 150 kHz to 200 kHz, an enormous increase over the 9 kHz allowed on medium and long waves! There is no hope of getting a lot of 200 kHz-wide stations on the 1000 kHz-wide m.w. band without considerable interference and only v.h.f. at present allows sufficient frequency 'space' or bandwidth for f.m. programmes to be used.

6.9 INTERFERENCE

As already stated, interference consists largely of high amplitude pulses of electromagnetic waves. Since the f.m. carrier is of constant amplitude, at the receiver a slicing or clipping circuit may be used to clip off any extra interference spikes, leaving the f.m. demodulator circuit to look for changes in frequency only. A well adjusted f.m. receiver will be almost immune to heavy amplitude-type interference, although poor settingup by some mechanics or technicians makes many customers doubt the advertised claims of 'interference-free reception'. Aiding the freedom from amplitude pulse interference is the fact that v.h.f. transmissions travel mainly via line-of-sight paths, so that it is unlikely that anyone in Britain will experience cochannel interference from a station using the same carrier frequency but situated in Europe.

6.10 PULSE MODULATION

Fig. 6.9 *a* shows a series of dots – it needs only a little imagination to 'see' a waveform instead of the dots. The dots could be the tips or peaks of pulses where an audio waveform is being sampled at regular intervals. Fig. 6.9 *b* shows such a system using a rotating switch mechanism that samples a speech channel, say 12 000 times a second for a very brief period per sample. At the receiving end there is a similar switch mechanism that rotates at exactly the same speed and exactly in step. These SYNCHRONISED switches connect the speaker to the listener so many times a second that the listener thinks that he is hearing continuous speech. In fact, the listener is possibly only

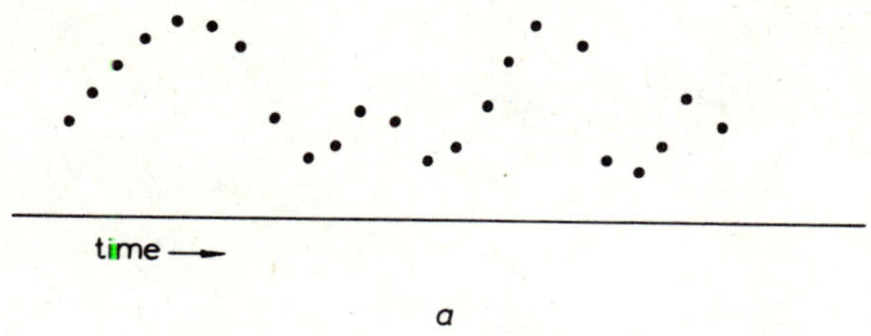

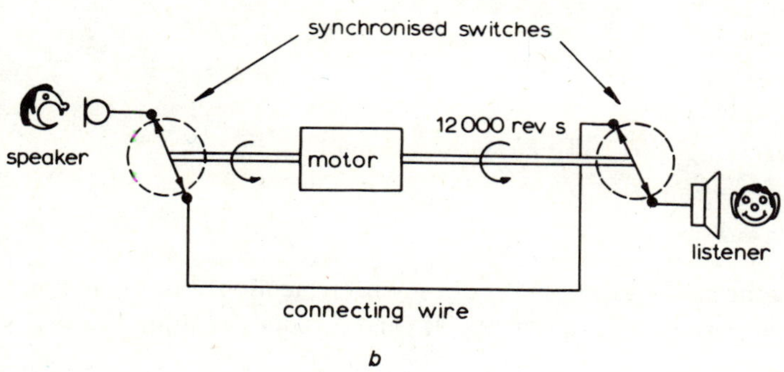

Fig. 6.9 Instantaneous voltages forming a waveform and a simple speech-sampling circuit

actually connected for one-tenth of the time, the rest of the time there being no connection, but the combination of the listener's hearing and the listener expecting to hear continuous speech fools the listener's brain.

The pulses of information may now be used to modulate a carrier. Either the a.m. or the f.m. system of carrier modulation could be used; an a.m.-type envelope is shown in Fig. 6.10.

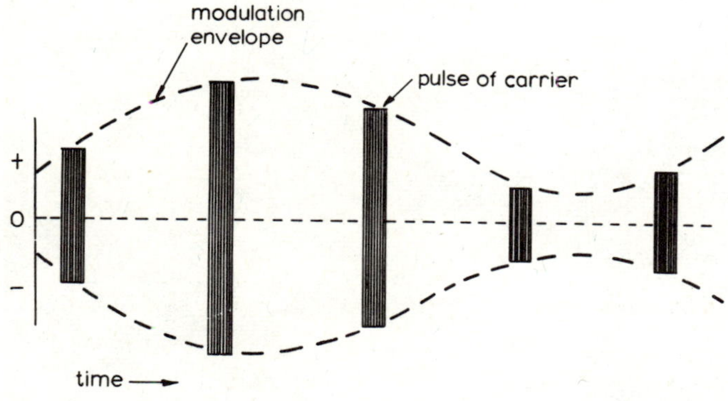

Fig. 6.10 Pulse-amplitude-modulated carrier envelope

The method is more complicated than either simple a.m. or f.m. systems since another set of audio processing circuits is called for to produce the specially separated pulses. If you look at this page of writing you will see that, apart from the actual words, there are great areas of white space. Between the lines and between the words is – nothing. Now listen to someone speaking; notice the gaps between words and imagine how long a time we spend actually silent when we are talking. A few more (carefully inserted) gaps are not noticed! And, in those gaps, other pulses may be inserted from other channels of audio as in Fig. 6.11 so that a single carrier (or wire in the instance of Fig. 6.9 *b*) may be used to convey many simultaneous information channels. The switching must be carefully synchronised of course or listener A will find that he is hearing speaker A, drifting to speaker B, drifting to speaker C and so on. This combination of several channels is called multichannel pulse modulation or TIME-DIVISION MULTIPLEXING, and while enalysed and proved to be workable in about 1935 it had to wait for transistors and integrated circuits before it became a feasible method. Using thermionic valves, a 10-channel system used about 100 to 200 valves, and spent much of its time being adjusted, usually due to valve ageing.

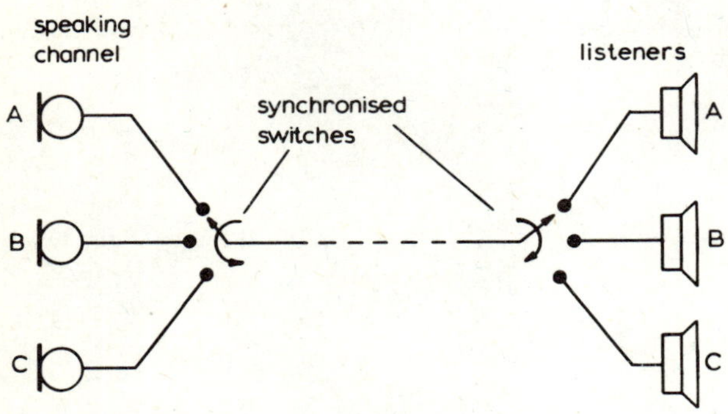

Fig. 6.11 Three-channel single-carrier system

6.11 PULSE MODULATION SYSTEMS

Although the method where the pulse tips follow the audio waveform has already been mentioned there are other variations; the three systems mostly used are (i) pulse amplitude (ii) pulse width and (iii) pulse position modulation.

(i) Pulse amplitude – already described – uses narrow pulses whose height is proportional to the audio waveform amplitude.
(ii) Pulse width – uses pulses whose width varies to more or less than the average pulse width, proportional to the audio waveform.
(iii) Pulse position – varies the position of a narrow pulse in time relative to a fixed position and proportional to the audio waveform.

All these three systems are shown for a ramp waveform 'audio' modulation in Fig. 6.12.

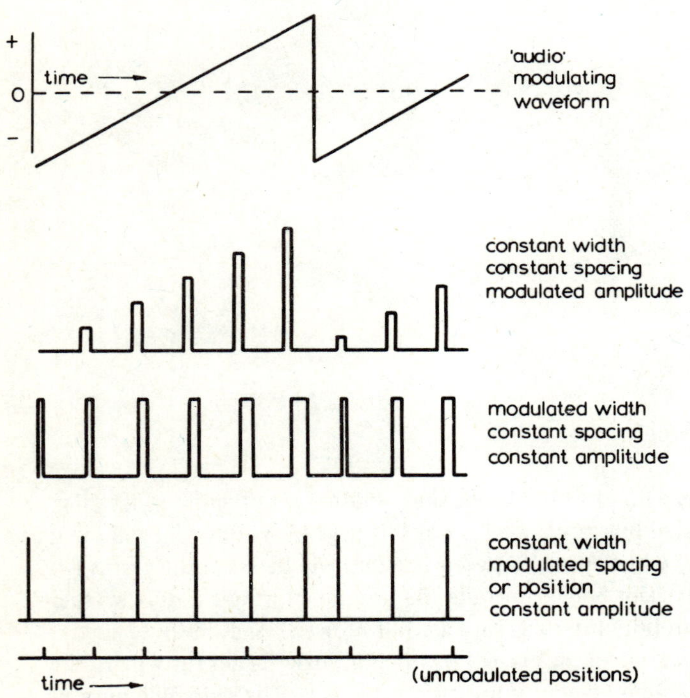

Fig. 6.12 Various pulse waveforms derived from same 'audio' modulation

Of the three variables possible, only one at a time is varied for each system – but now imagine the permutations possible if multichannel working for, say 10 channels is used with pulse height modulation; a further 10 channels added using the same pulses and using pulse width modulation, and then a further 10 channels added to the same pulses using pulse position modulation. What a waveform!

6.12 SYNCHRONISATION

A special pulse – or a special gap between pulses is usually inserted in the final signal. At the receiver this special pulse – or gap – is identified and used to synchronise the receiver electronic switching with the transmitter channel switching. A very similar system is in use for all t.v. receivers to keep the horizontal- and vertical-scanning systems exactly in step and to stop the picture rolling or slipping.

6.13 BANDWIDTH FOR P.M.

Pulse modulation systems require a large bandwidth; a true square-topped pulse involves frequencies up to infinity, so that modulation, whether a.m. or f.m. onto a carrier is going to need very large bandwidths. In fact, the pulses transmitted are not perfectly square but even so, the bandwidth after modulation is still large and only suited to v.h.f. or u.h.f. where there is at present sufficient frequency 'space' available.

Chapter 7

Demodulation

7.1 INTRODUCTION

At the receiver the wanted signal or programme is selected and then the modulating signal is to be separated from the carrier, i.e. the received carrier is to be DEMODULATED.

7.2 THE 'CRYSTAL SET' RECEIVER

Demodulation invariably involves the use of DIODES that are one-way conductors sensitive to the polarity of an applied signal. Fig. 7.1 shows the circuit of a simple crystal set. When broadcasting first began about 1925, the ordinary person could not afford a 'valve', so little pieces of various mineral crystals about 5 mm across were probed or 'tickled' with a finely sprung copper wire – the 'cat's whisker' – until the correct spot was found and the programme was faintly heard in the earphones.

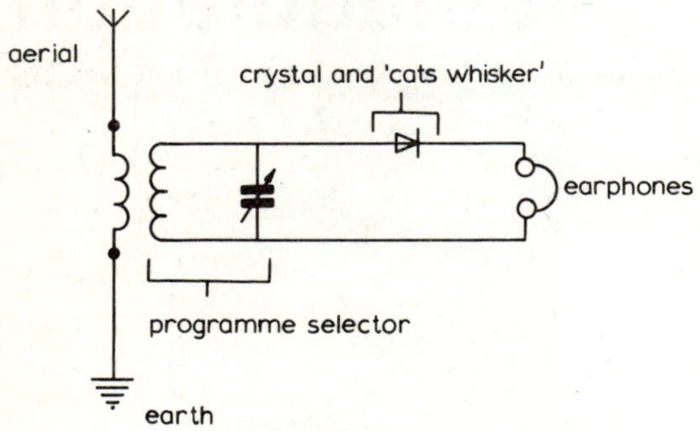

Fig. 7.1 Crystal-set receiver circuit

Nowadays the crystal and cat's whisker are replaced by a 'detector diode', a modern semiconductor equivalent. The coil of 40 or 50 turns wound on a cardboard tube some 5–10 cm diameter and tuned by a variable capacitor are still very similar to the original components and high-resistance earphones then complete the circuit.

The signal induced into the tuned circuit is of a small voltage swinging above and below zero at radio frequency and the amplitude of these swings is varying in sympathy with the audio programme at the transmitter. The detector diode allows current to flow every time the voltage at the top of the coil swings positive, but the diode stops conducting when the voltage swings negative.

Through the earphones there is a series of pulses of r.f., the height, size or amplitude of these pulses varying at an audio rate as in Fig. 7.2*b*, this waveform having been obtained from the amplitude modulated carrier waveform shown in Fig. 7.2*a*. The earphones, which are electrical-to-sound energy transducers, respond to the audio variations of current by reproducing sound.

It is easy to build one of these simple receivers – but *do* use a long aerial of 20 to 50 metres *and* a good earth system if you wish for any more than a very faint sound. At least you will not have to worry if someone walks across the room, the resulting vibration dislodging the cat's whisker from the crystal's sensitive detecting spot that you had taken half-an-hour to find – modern detector diodes are built to last longer than that!

7.3 THE NONLINEAR DIODE ACTION

Leaving the practical circuit and returning to systems, the diode is known as a 'nonlinear' element; in other words the way that it behaves depends upon the polarity (and sometimes the amplitude) of the applied voltage or current. A simple resistor is a 'linear' element; it behaves the same whether fed with d.c., a.c., a.f., or r.f.

Now any waveform applied to a nonlinear device or element is distorted at the output. If a pure sine wave – a single frequency – voltage is applied, a halfwave rectified waveform is the result and it has already been stated that any non-sine wave contains a basic frequency as well as harmonics. Thus a nonlinear device distorts and produces frequencies in addition to the original frequencies.

7.4 MODULATION PROCESS

The modulating stage at the transmitter is a nonlinear stage into which is fed simultaneously the carrier and modulating signals, radio and audio frequencies. The result at the output is the presence of the original two frequencies and the side frequencies that go to make up the modulated carrier. The original audio signal is not 'transmitted' since it is not a radio frequency and cannot radiate in the same way from the aerial which is acting as a form of filter.

7.5 DEMODULATION

At the receiver the carrier and side frequencies are applied to a nonlinear device – the crystal in Fig. 7.1 – and further distorted. Fig. 7.2 *b* is a distorted form of Fig. 7.2 *a* but it is seen that the detected (and distorted) signal is now an audio waveform 'sitting' on a d.c. level and able to operate a pair of earphones. The original modulated carrier contained the information but could not affect earphones; being a *radio* frequency.

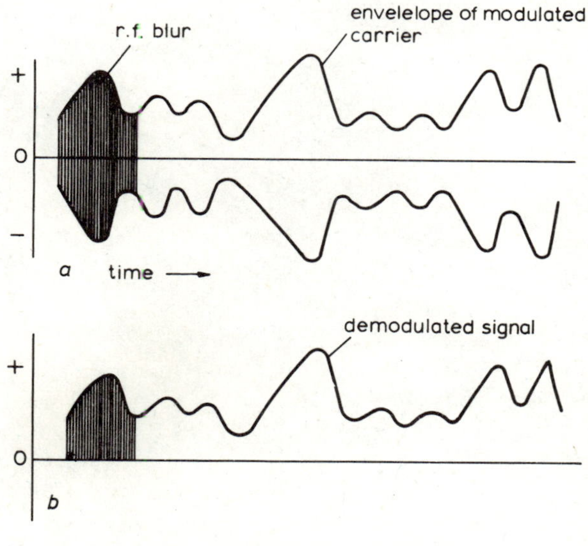

Fig. 7.2 A.M. signal before and after demodulation by diode

7.6 MIXING

If two frequencies – not specifically radio or audio frequencies this time – are applied to a nonlinear device, the output contains the original two frequencies PLUS their harmonics PLUS the sum(s) PLUS the difference(s) of the original frequencies.

7.7 MULTIPLICATION

Even if only one frequency is applied to a nonlinear device, the distorted output contains harmonics, i.e., two, three, four (and higher) times the original frequency. In other words, the frequency has been MULTIPLIED and the wanted harmonic may be then selected and used.

7.8 GENERAL EFFECT OF NONLINEAR DEVICES

In general, in Fig. 7.3, if the INPUT is of frequency f_A, the OUTPUT will contain f_A, $2f_A$, $3f_A$, $4f_A$ . . . etc. With frequency f_B applied the output contains f_B, $2f_B$, $3f_B$, $4f_B$ . . . etc. And with both f_A and f_B applied at the same time, the output contains the frequencies and harmonics already listed above as well as $f_A + f_B$, $f_A - f_B$, $2f_A + f_B$ and so on – so many different frequencies that almost any simple mathematical relationship between f_A and f_B may be found in the output frequency, and selected with the aid of suitable filters.

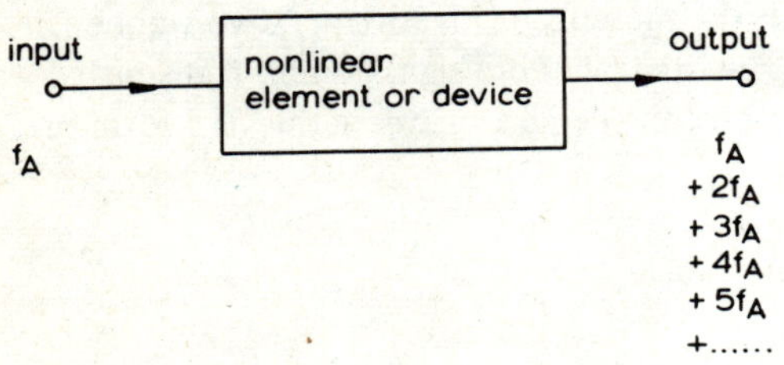

Fig. 7.3 Input frequency 'distorted' and multiplied

So, although given different names, the 'blocks' of modulator, mixer, demodulator, detector and frequency multiplier that you will come across in diagrams are very often similar in their processing of applied frequencies – deliberate waveform distortion and harmonic production.

Chapter 8

The amplifier

8.1 AMPLIFIER SYMBOLS

An amplifier contains a minimum of a single transistor or valve – and nowadays a maximum of hundreds of transistors in an integrated circuit. The function of an amplifier is to make the input signal larger at the output and for this output signal ideally to be an undistorted replica of the input signal, unless the amplifier is a special type.

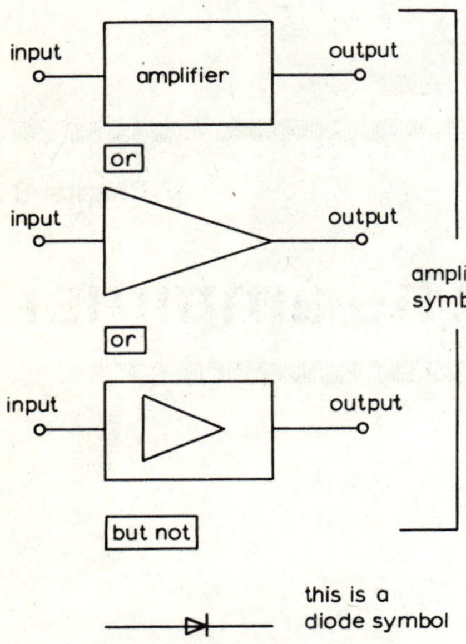

Fig. 8.1 Amplifier symbols

The amplifier may appear as a labelled block or as a triangle or as both these, the triangle within the block; see Fig. 8.1. The direction of the arrowhead indicates signal flow, from input→output or from small signal→larger signal; the symbol must not be confused with that for a rectifier which is usually smaller and is drawn slightly differently.

8.2 PHASE RELATIONSHIPS

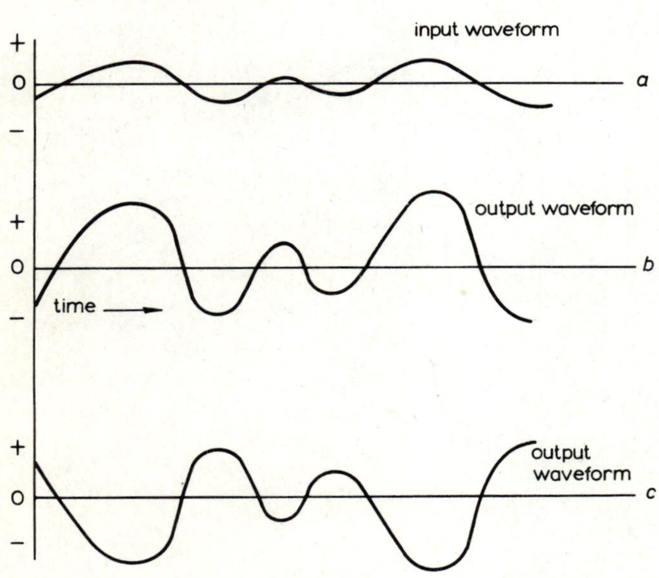

Fig. 8.2 Amplifier input (*a*) is larger in (*b*) but of the same phase, while (*c*) has been phase reversed

Although the waveforms at input and output terminals are differently sized, this is not usually indicated on a block diagram; the amplifier symbol indicating that the output is the larger amplitude. What is more important is the PHASE relationship between input and output waveforms. In Fig. 8.2 are shown typical input and output voltage waveforms. If Fig. 8.2 *a* has its positive – and negative – going peaks compared with Fig. 8.2 *b* and then Fig. 8.2 *c* it is seen that Fig. 8.2 *a* and Fig. 8.2 *b* are exactly in step or in phase, whereas Fig. 8.2 *a* and Fig. 8.2 *c* are exactly out of step or phase-reversed.

The complete amplifier block is often made up of several transistors or valves and these are called 'stages'. As a rule, each single stage results in one complete phase reversal so that two stages in cascade will give two reversals and bring the output back into step or in phase with the input once more. The phase relationship is often shown by 'steps' or 'top-hats' as in Fig. 8.3, where three stages in cascade are shown. The use of these square-edged waveforms is only to show phasing, not to suggest that the actual signal is of that shape, and it is seen that the output is of reversed phase compared with the input waveform.

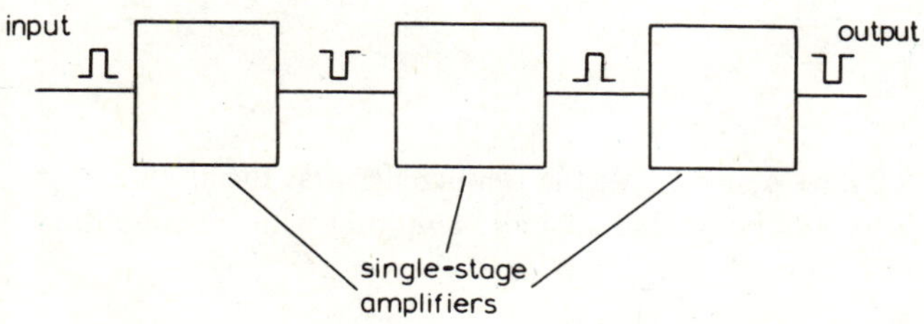

Fig. 8.3 Cascaded amplifiers

The importance of the phase relationship between amplifier input and output will be seen later when feedback is discussed.

8.3 AMPLIFIER TYPES

The voltage amplifier is designed mainly to accept a small voltage at the input and to provide a larger voltage of the same waveform at the output. Since the input terminals do not look into an absolute open circuit, within the amplifier block a small current must flow from the signal source. At the output, the amplified voltage will also feed a current into the load, even if only a voltmeter or another amplifier stage is acting as the load, and so the voltage amplifier must also act as a current amplifier at the same time.

A current amplifier will be designed to accept small currents and to provide larger currents at the output; as argued above, there must also be voltage present at input and output so this is also a voltage amplifier.

A power amplifier is one which is designed to produce a considerable voltage *and* current swing at the same time, i.e., plenty of output power to feed a load like a loudspeaker, aerial or perhaps motor.

A typical record player uses a crystal pickup or transducer feeding small voltage variations to a two-stage voltage amplifier, which in turn feeds a power amplifier stage and this operates the loudspeaker transducer as in Fig. 8.4.

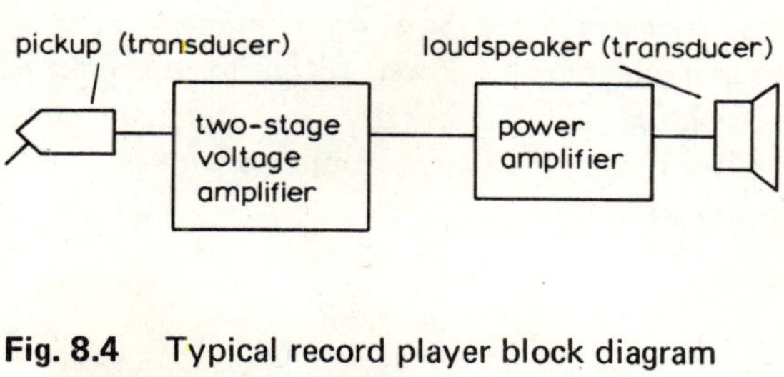

Fig. 8.4 Typical record player block diagram

A single output stage using one transistor or valve is rather inefficient and a PUSH-PULL output stage is often fitted instead. The usual type of push-pull output stage needs two input signals, one for each stage, and these signals need to be the same amplitude but of opposite phase so that as one output stage 'pushes', the other 'pulls' and vice versa. The DRIVER stage providing these two out-of-phase signals is usually called the PHASE-SPLITTER stage, and Fig. 8.5 gives a block diagram showing signal phasing, the outputs being combined, in this instance, in a transformer.

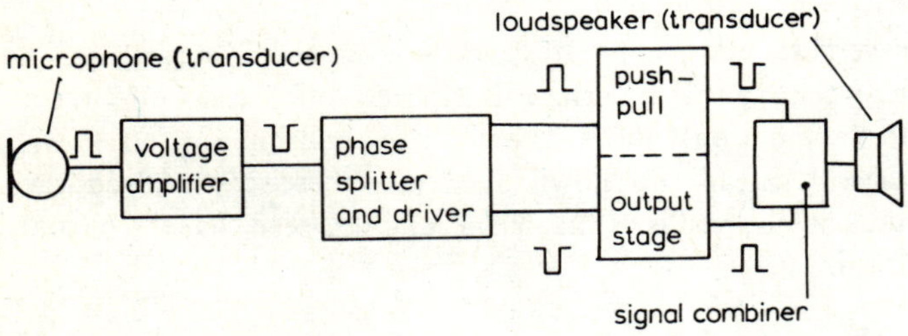

Fig. 8.5 Amplifier with push-pull output stage

The components used within an amplifier stage or block have control over the frequencies that the amplifier will accept. An amplifier that handles from 20 to 20 000 Hz would be termed an audio amplifier while an amplifier that handled say 5 to 8 MHz, would be an r.f. amplifier.

A special title is used for the amplifier that feeds the brilliance modulation information to the cathode ray tube in a t.v. receiver; the frequencies involved cover both a.f., and r.f., going from about 5 Hz to 5 MHz; a VIDEO amplifier is needed for this range of frequencies.

Another special amplifier title – actually a low-gain voltage type as a rule – is the BUFFER amplifier. This is used after an oscillator whose frequency or voltage output might be affected by connecting directly to a load or normal amplifier. Used like the buffers between train wagons, the buffer amplifier possibly gives little amplification but serves mainly to stop the two other stages from affecting each other.

8.4 EFFICIENCY

All amplifiers consume energy and need battery or power supplies to keep them running. An amplifier that is doing nothing – switched on but with no input signal – is of course 100% inefficient or 0% efficient. A voltage amplifier is also fairly inefficient in terms of its Power output/Power input ratio but it is usually with power amplifiers that efficiency becomes important.

A power amplifier might run at full output with 3 W available to the load; if the power taken from the d.c. power supply is 6 W then the efficiency is: Audio Power Out/DC Power In $= 3/6 \times 100\% = 50\%$. The main reason for using push-pull output amplifiers is the increased efficiency that it is possible to obtain, typically 60% for push-pull compared with 30% for a single-ended output stage. If you are carrying around equipment then the battery size and weight becomes important, even more so where you have to keep buying more batteries for an inefficient apparatus that seems to produce little useful output for a large consumption.

Valved apparatus is automatically less efficient than transistorised apparatus since the power required to run the valve heaters is not taken into account in efficiency calculations as a rule, but this power still has to be supplied to keep the valves in working condition. With about 6·3 V at 0·3 A being the typical heater power requirements, i.e., 1·9 W per valve, it is seen that apparatus using four or five valves is most inefficient compared with similar transistorised circuitry that does not need the extra 10 W heating power.

8.5 FREQUENCY RESPONSE – BANDWIDTH

As already mentioned, amplifiers have their characteristics controlled by the choice of particular components used, and it is often necessary to find out just how low or how high a frequency the amplifier will amplify adequately. This information may be needed after construction, after repair or servicing, or just to check that the designer's specifications are correct after purchase of the apparatus.

Considering an audio amplifier, a quick check would be to connect a load – probably a loudspeaker – and a microphone to the amplifier, speak or sing into the microphone and listen to the resulting sound from the speaker. But it is difficult to speak and listen at the same time, while a fellow worker will not be able to tell much about the actual frequencies in use by just listening.

The next step is to replace the audio input, whether microphone or record pickup, by an audio frequency generator. This will feed in known frequencies, usually a pure sine wave, and also allow the input voltage (or current) to be measured easily since it is now a steady value. To feed the output to a loudspeaker and rely on the ear will be a good test from the point of view that the amplifier is tested under 'listening conditions', but has every listener's ear the same characteristics and high and low frequency sensitivity? And also, testing a high power amplifier for half-an-hour will become very wearing on nerves and hearing, and even more so on the rest of the staff in the workshop and neighbourhood.

The solution is to use a 'DUMMY LOAD' that has the same resistance or loading effect as the normal load or loudspeaker and yet produces no noise. A high power resistor is usually used with a voltmeter or ammeter to show the voltage, current and power being developed. You probably have already learnt that for a pure resistor,

$$(1)\ \text{Power} = \text{voltage} \times \text{current} = VI\ \text{watts}$$
$$\text{or}\ (2)\ \text{Power} = I^2R\ \text{watts}$$
$$\text{or}\ (3)\ \text{Power} = V^2/R\ \text{watts}$$

so that a voltmeter and known value of load will enable the output power to be calculated if required. A further requirement is that the instruments used should be able to respond adequately to the frequency in use – not all a.c. instruments will read correctly at 15 Hz or 10–20 kHz – and this should be checked as well.

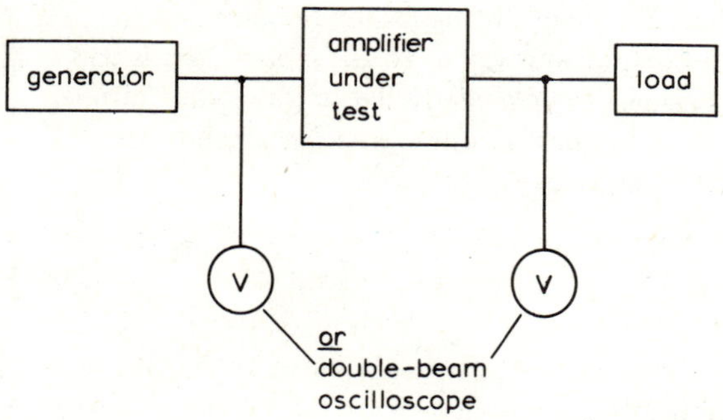

Fig. 8.6 Amplifier testing setup

Fig. 8.6 shows the frequency response test setup; it is very convenient to use a double-beam scope if this is available since the waveforms and phase relationships and voltages are all seen at the same time, otherwise a.c. millivoltmeters and voltmeters may be used.

The input voltage should be kept small and of constant value or level so that the output is not too large and becomes distorted because of overdriving (this is where the scope is so useful, to see the onset of distortion). The generator frequency is varied in suitable steps – note the frequency each time – the input voltage is checked to be the constant value decided upon, probably between 1 and 10 mV and the output voltage is also noted.

Looking at the test figures will show roughly at which frequencies the response of the amplifier is falling off but a better display of results is in Fig. 8.7 where the FREQUENCY RESPONSE curve is plotted. The horizontal base of frequency is to a logarithmic scale as a rule, where 10–100 Hz requires the same length as 100–1000 Hz, 1000–10 000 Hz and so on.

The x^2 or x^3 scale of your slide rule is probably the easiest way to calibrate the frequency scale on the paper since logarithmic scaled graph paper is not so common and is more expensive than normal 'linear' graph paper.

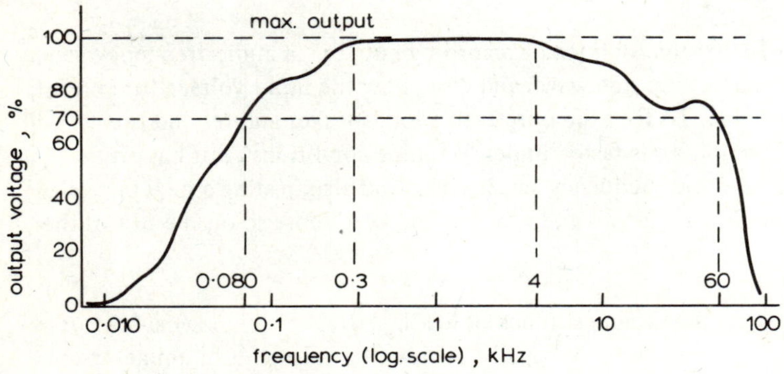

Fig. 8.7 Frequency response of an amplifier

Although the graph shows that the 'flat response' is from 300 Hz to 4 kHz, the BANDWIDTH is normally taken between points where the curve has dropped to 0·7 (0·707 to be accurate) of the maximum or 'flat' output. The graph shows that the response of the amplifier tested is from 80 Hz to 60 kHz between these '0·7 times maximum' points, i.e. a total bandwidth of 60 minus 0·08 = 59·92 kHz.

The reason for using these '0·7 times maximum' points (or if you wish to be technical, '3 dB down') is that the final receiver in use is invariably the ear and the ear is just about capable of distinguishing the fact that the power output is less when the voltage to the transducer drops to 0·7 of its previous value. So as far as the ear is concerned, the bandwidth as specified is what the ear and brain considers to be a nearly flat or unvarying output, even though instruments show considerable change. As a matter of interest, this amplifier has a bandwidth extending well beyond the capabilities of the human ear and of most loudspeakers, too.

8.6 DISTORTION

This is a word that has many meanings, not all of them the same; it does not necessarily mean unpleasant or rough sound, it *does* mean for our purposes that an output waveform is different in shape from the input waveform in an electronic system. Thus if the voice of a speaker is faithfully converted to voltage variations by a microphone, and the amplifier and loudspeaker then faithfully reproduce the voice – possibly louder and a long way away – there is no distortion. If however the tone controls are adjusted to make the voice sound 'better', or the loudspeaker is in a large 'boomy' hall and the bass response is reduced to make the voice more intelligible, then either of these adjustments causes distortion of the original input signal, the voice.

Equally well, the process of modulation produces a distorted waveform and the demodulation process also distorts, the hope being that in the end, the original signal will be obtained with as little overall distortion as possible. It is seen that 'distortion' is often a requirement of a piece of equipment or of one block within the complete block diagram, and the final test of distortion is whether the complete channel of information fulfills its function of allowing communication –poor, reasonable or excellent – from input to output.

Suppose that an amplifier has a frequency response from 300 to 800 Hz as in Fig. 8.8. If a sine wave of 500 Hz is applied to the input then a sine wave of 500 Hz would be available at the output, thirty times larger since the gain is specified as '× 30'. If, however, a nonsine wave of 500 Hz is applied, then the output will be almost a pure sine wave – a distorted output – since the input signal must have contained 500 Hz and some of the range of

2 x, 3 x, 4 x, (etc.) 500 Hz, the harmonics. Since even the second harmonic at 1000 Hz is barely amplified by the amplifier (whose response curve shows that a 1 kHz input is only three times as large at the output as is the same amplitude of 500 Hz input) then only the fundamental of 500 Hz is passed through the amplifier and any harmonics that were present at the input to make up the particular shape of the input waveform are attenuated or filtered out by this amplifier.

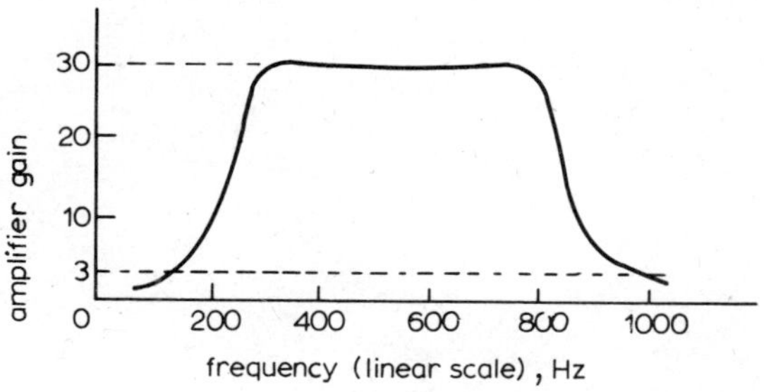

Fig. 8.8 Frequency response of narrow-band amplifier

With the same amplifier and a 500 Hz sine wave input, if the input voltage is increased to too high a level, then the output voltage will be unable always to be 30 times larger and the tops and bottoms of the output waveform tend to be flattened. This is because there is a limit to the ability of the output stage to swing the voltage up and down, usually set by the d.c. supply rail voltages in use. Once again distortion occurs – even though the input frequency is within the response curve limits – because the output waveform is not the same as the input waveform.

Chapter 9

Feedback

9.1 INTRODUCTION

So far an amplifier has been considered to be fed from an alternating source and then to feed the larger output to a load of some form. Suppose that at the same time as supplying the load, some of the output is taken round to combine with the input as in Fig. 9.1; there is FEEDBACK.

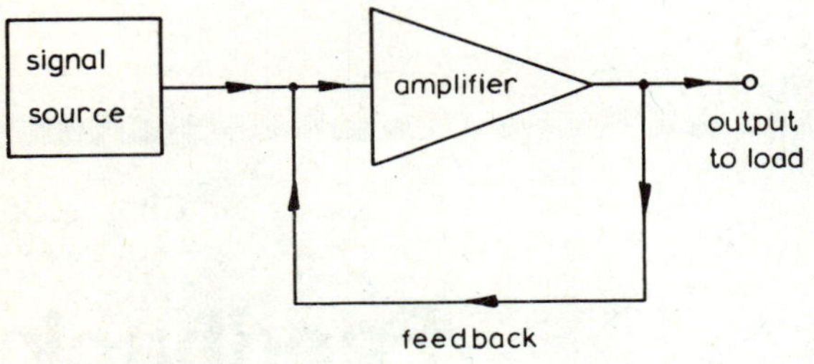

Fig. 9.1 Amplifier with feedback

There are two types of feedback, negative feedback (n.f.b.), or positive feedback (not usually abbreviated) and the type is decided in the following way:

If the output phase is the same as the input phase then the feedback signal will combine with the input signal to make it larger than before, Fig. 9.2, giving POSITIVE FEEDBACK.

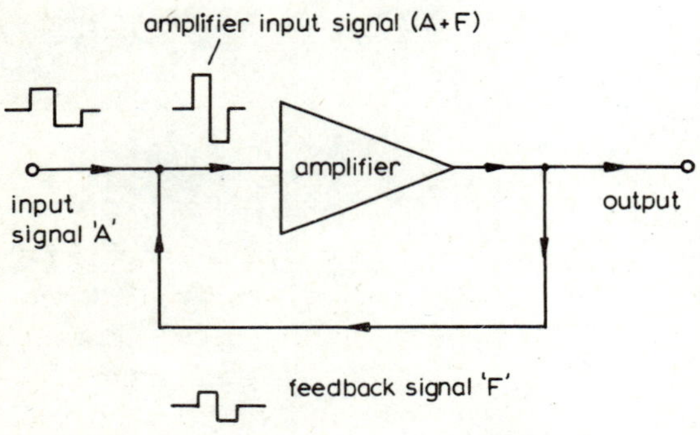

Fig. 9.2 Amplifier with positive feedback

If the output phase is the opposite of the input phase, then the feedback signal combines with the input signal to make it smaller than before as in Fig. 9.3, giving NEGATIVE FEEDBACK.

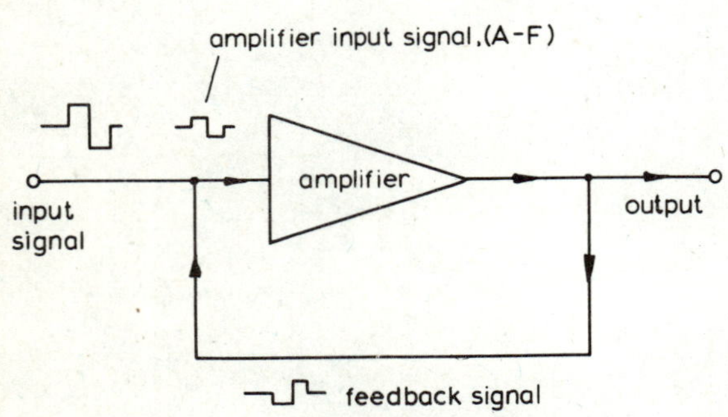

Fig. 9.3 Amplifier with negative feedback

9.2 NEGATIVE FEEDBACK

An argument must be used (in the absence of maths) to demonstrate the effects. Imagine the amplifier to give a gain of one hundred times to the input voltage, 1 V being fed to the input terminals. Without feedback the output will be 100 V as in Fig. 9.4 *a*. Now arrange feedback – negative – of one-tenth of the output signal, i.e. 10 V.

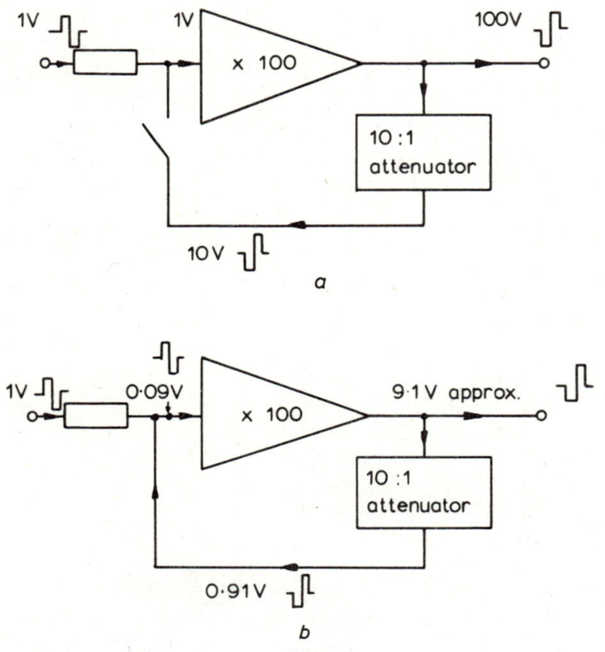

Fig. 9.4 Amplifier with one-tenth of output as n.f.b.

The 1 V input with 10 V of the opposite phase added would result in more than just cancelling the original 1 V, it would leave 9 V of the output phase at the input. But, if all of the input phase voltage is cancelled there will not be any of the output phase that we are using as the negative feedback signal, so the result of this struggle for supremacy at the input is that the original input phase is considerably reduced in amplitude, then amplified by the constant gain of one hundred, but giving much less than 100 V out and then one-tenth of this smaller output being fed back to reduce (but not cancel) the 1 V input down to its particular value. It is a balancing effect and the actual output voltage probably ends up at about 9 V or so, as shown in Fig. 9.4 *b*.

9.3 EFFECTS OF NEGATIVE FEEDBACK (N.F.B.)

If the amplifier is less than perfect it produces some distortion that is added at the output to the amplified input signal and perhaps also adds in some hum from the power supply and noise from the valves and transistors. The output thus consists of signal and distortion products + hum + noise as in Fig. 9.5 *a*. Only the original signal was fed to the input and the addition of negative feedback now results in distortion products + hum + noise being deliberately introduced at the input so that they will try to cancel themselves by being amplified times, say one hundred, and arriving at the output in opposite phase to the input – which was the output. Fig. 9.5 *b* shows the final result.

Unfortunately, the cancellation is not perfect, since, if there were finally zero noise, hum or distortion at the output there would be zero feedback of noise, hum or distortion; no input of these unwanted products – and hence, no cancellation of the noise, hum and distortion produced within the amplifier.

(*a*) N.F.B. REDUCES the effective overall GAIN from input to output.
(*b*) N.F.B. REDUCES the unwanted NOISE at the output.
(*c*) N.F.B. REDUCES the unwanted HUM at the output.
(*d*) N.F.B. REDUCES the unwanted DISTORTION products.

These last three reductions are most worthwhile and result in the amplifier behaving much better. It is a pity that the overall gain is reduced as well, but it is easy to design-in far too much gain so that after n.f.b. has been adjusted to reduce the unwanted noise, hum and distortion by-products the reduced effective gain of the amplifier ends up as the desired value.

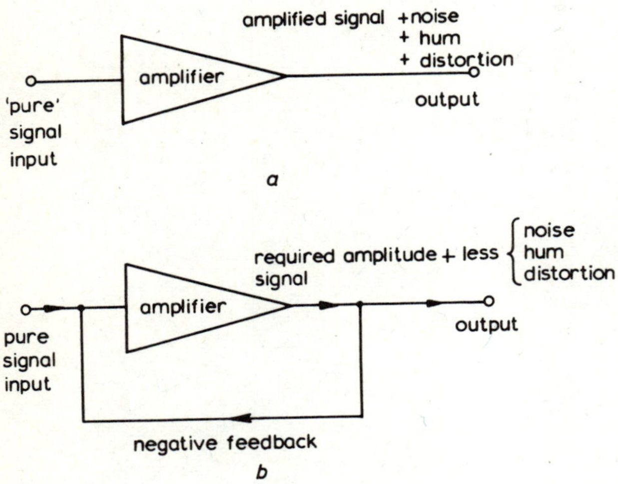

Fig. 9.5 Comparison of amplifiers with and without n.f.b.

The feedback line may have filter circuits added so that selected frequencies only are fed back reducing the gain of the amplifier for these selected frequencies but allowing all other frequencies to see the amplifier as a full-gain device. If this is done then the overall effect is of 'tone control' and it is quite common in an amplifier circuit to be able to trace through three or four feedback circuits 'designed in' to control various bands of frequencies.

9.4 POSITIVE FEEDBACK

Positive feedback is exactly opposite to n.f.b., and the effects of positive feedback are opposite as well. The effective gain is increased and so is distortion. Positive feedback cannot introduce hum and noise if none were there before, but it will certainly make the effects of hum and noise worse rather than better.

Used very carefully and in a controlled way an amplifier's gain may be increased by small amounts of positive feedback. If a little too much positive feedback is used however, then any variation of the output will feed back enough to the input – amplified – to provide a larger output to feed back enough to the input – amplified – to provide an even larger output . . . the amplifier is now quite satisfied without any particular input source because the output provides the input which provides the output.

Positive feedback usually produces OSCILLATION where the amplifier (plus positive feedback) may be considered as an a.c. generator needing only a supply of power to keep it running. All the a.f. and r.f. generators that one sees or uses are oscillators and use positive feedback; every t.v. receiver uses at least three oscillators and cannot function without them.

9.5 THE OSCILLATOR

An oscillator is an amplifier with deliberate provision of positive feedback, often through a selective network that determines the frequency and waveform of oscillation available at the output, Fig. 9.6*a*.

If the feedback network selects or filters one frequency only and the amplifier has just enough gain to maintain oscillation without distorting the frequency that is fed around the circuit, the output will be a sine wave.

If the gain is too large, distorting whatever the feedback network allows back from output to input, or if the network is not particularly frequency selective then the output will be a complex waveform and not sinusoidal; see Fig. 9.6 *b*, for typical output waveforms.

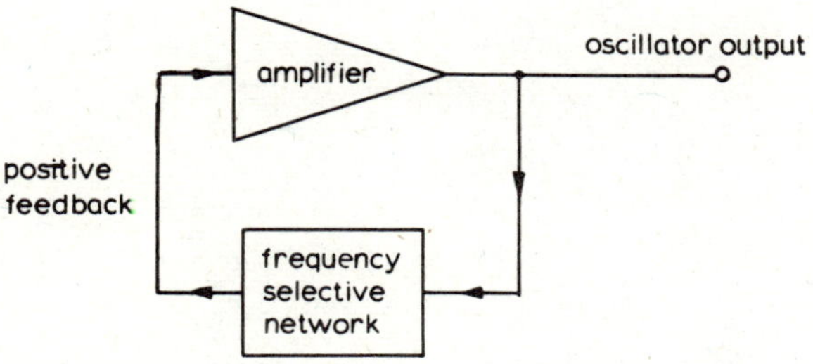

a

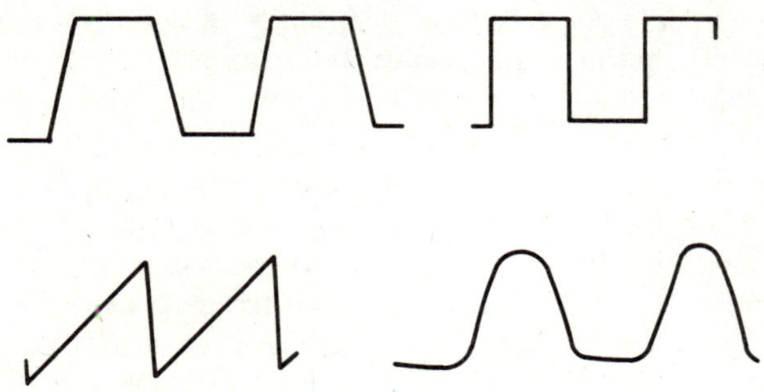

b

Fig. 9.6 *a* Positive feedback amplifier forming an oscillator
b Some typical output waveforms

An oscillator is often followed by a buffer amplifier as in Fig. 9.7, so that the load fed by the output is isolated from, and has no effect upon the oscillator amplifier-feedback system, thus not affecting the frequency or waveform of the output from the oscillator itself, even though the output from the buffer amplifier may be affected by the load.

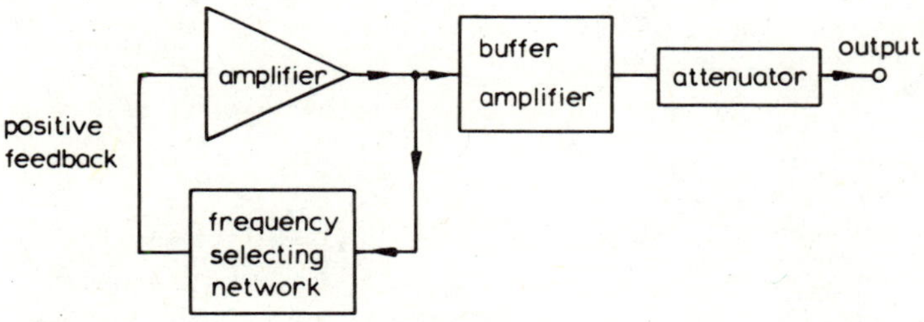

Fig. 9.7 Block diagram of a generator

9.6 AUTOMATIC GAIN CONTROL (A.G.C.)

An amplifier can be designed so that its gain may be electronically controlled rather than need to have the shaft of a gain control moved by hand.

Feedback from the output end of Fig. 9.8, via a special network to a point near the input end of the amplifier is used to control the gain automatically.

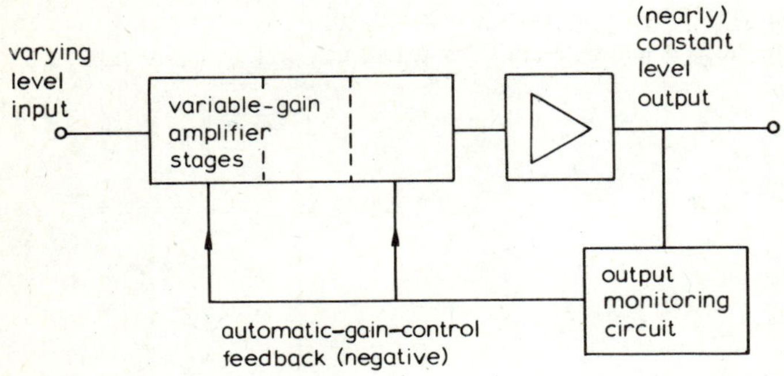

Fig. 9.8 Use of a.g.c. to maintain constant output level

The words 'automatic gain control' normally refer to the system used in radio receivers where a.g.c. feedback is used to counteract the effects of distant stations fading in strength. A.G.C. also cuts back the gain of strong local signals compared with distant weak signals, making the output level more nearly independent of station locality. A.G.C. used to be called a.v.c. – automatic volume control – but the work 'gain' is preferred since the a.g.c. process actually controls stage gain.

A similar process, sometimes called automatic level control or a.l.c., is used in many transmitters to vary the microphone amplifier gain in such a fashion that the modulation level is kept near maximum, regardless of whether the operator is speaking quietly or nearly shouting into the microphone. This process is very necessary for apparatus that has the bare minimum of controls, probably little more than a transmit/receive switch. The modern cassette or tape recorder often has 'automatic record level' circuitry fitted for the benefit of the user who no longer has to keep watching a flickering magic eye or instrument needle while recording.

9.7 AUTOMATIC FREQUENCY CONTROL (A.F.C.)

Oscillators are used in superheterodyne (or superhet) radio receivers, and, particularly in v.h.f. types, the oscillator is liable to drift and slowly shift the programme being received out-of-tune. Once correctly tuned in an a.f.c. circuit may be switched in whereby a feedback voltage is fed back from a frequency comparator circuit to the oscillator as in Fig. 9.9. If the oscillator drifts, the comparator notices the change of frequency, produces a d.c. correction voltage, and almost corrects the drift to maintain 'on tune' conditions for hours.

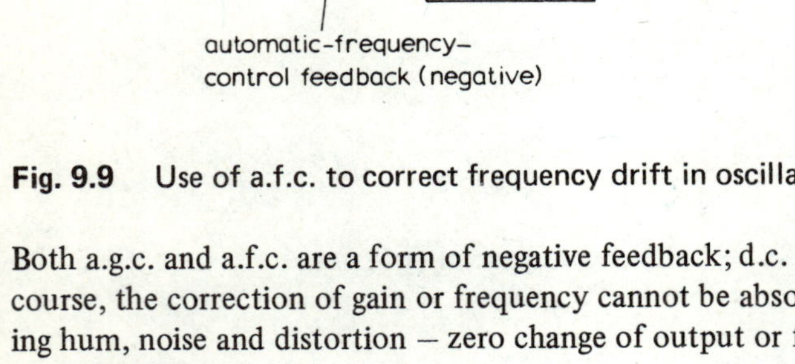

Fig. 9.9 Use of a.f.c. to correct frequency drift in oscillator

Both a.g.c. and a.f.c. are a form of negative feedback; d.c. is used as the n.f.b. signal in both instances and, of course, the correction of gain or frequency cannot be absolutely perfect since – as discussed with n.f.b. correcting hum, noise and distortion – zero change of output or frequency at the output end would produce a zero correction signal.

Chapter 10

Coupling circuits

10.1 INTRODUCTION

The blocks in block diagrams are coupled together quite simply by a single line along which the signal, waveform or voltage is passed. In a real circuit however, there are invariably two lines – go and return, – one of which is often the chassis or earth connection. And, the coupling circuit itself, though quite simple and involving only two components, is able to reshape a waveform quite considerably.

10.2 CAPACITOR COUPLING, D.C. BLOCKING, SINE-WAVE INPUT

One of the most common needs in electronics is to be able to couple a small alternating voltage 'sitting' or superimposed on a direct voltage at one place across to another point in the circuit where the direct voltage is considerably different. This is illustrated in Fig. 10.1 *a* where at A the collector voltage of TR1 is, say 8 V d.c. relative to the chassis while at B the base voltage of TR2 is 0·5 V d.c. relative to the chassis.

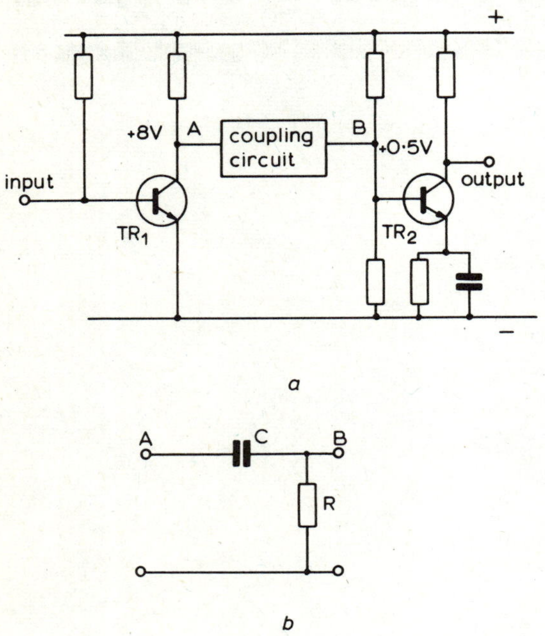

Fig. 10.1 *a* Showing need for d.c. isolation between A and B and
b A *CR* coupling circuit

TR1 and TR2 are amplifiers and the output waveform at A needs to be coupled to the input of TR2 at B yet a direct join will (i) rob TR1 of its working voltage or (ii) probably damage TR2 with excess voltage.

A capacitor is an open circuit to a flow of d.c. and yet offers low opposition to a flow of a.c., the opposition depending upon the frequency of the a.c. and the value of the capacitor. By connecting a suitable value of capacitor from A to B the alternating waveform is coupled across but there can be no flow of d.c. – just as required. The actual value of the capacitor will not be determined at this stage, but a general rule is that a sine wave output at A will be coupled across to B and appear as a sine wave, although possibly a little smaller and shifted in phase.

The complete coupling circuit is considered to be both the capacitor and the following resistor (which may have to take into account the effective resistances in parallel with the actual resistor) so that it is often referred to as a CAPACITOR-RESISTOR or a *CR* coupling circuit and shown as Fig. 10.1 *b*.

10.3 *CR* COUPLING OF SQUARE WAVE – DIFFERENTIATING

If a similar *CR* circuit is used to couple a square wave from the output of one stage to the input of the next stage then the shape will be considerably changed. The square wave is the result of adding together many different sine waves of increasing frequency and each of these is attenuated and shifted in phase by the *CR* circuit so that when these are added together the result is a spiky waveform. This *CR* circuit is called a DIFFERENTIATING CIRCUIT because its output is the rate of change of the input, a feature often required in electronic circuits.

Another way of looking at this circuit is to remember that the capacitor will couple across the change of waveform and yet be open circuit to d.c. The vertical portions of the square wave represent changes of voltage but the horizontal portions are just steady values or small intervals of direct voltage. So, when the input suddenly changes there is a rate of change and the output changes. When the input is steady there is no rate of change and the output side of the capacitor will return to zero by drifting back at a rate determined by the product $C \times R$ of the circuit. This product (C in FARADS and R in OHMS) is called the time constant and is measured in seconds.

For a given *CR* circuit, it is not possible to tell, without calculations, whether the circuit is being used for straightforward coupling and d.c. blocking or as a differentiating circuit. Four output wave forms shown in Fig. 10.2 *b* to *e* indicate how various increasing *CR* values affect a square input waveform and, in particular, Fig. 10.2 *e* shows that a nondistorting coupling arrangement has nearly been reached with a particular combination of *C* and *R*.

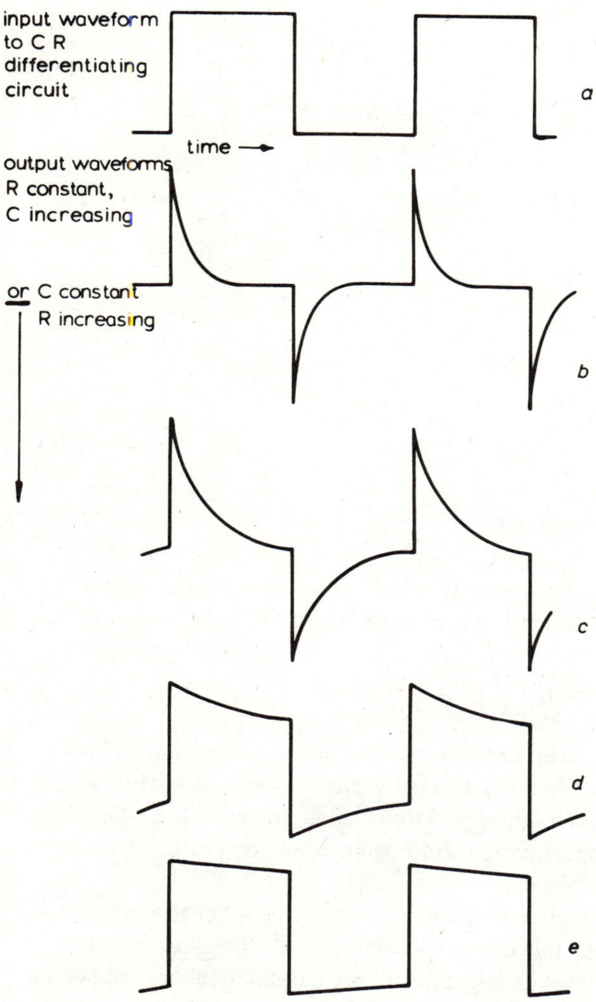

Fig. 10.2 The result of feeding a square wave into a *CR* differentiating circuit

10.4 RESISTOR COUPLING

When it is not important to block the flow of d.c. but merely to limit the value of current, a resistor may be used. The coupling resistor will probably be feeding into another circuit – the input of the next stage – and may therefore be redrawn from Fig. 10.3 *a* through Fig. 10.3 *b* to an equivalent in Fig. 10.3 *c* where the two resistors are seen to form a POTENTIAL DIVIDER circuit.

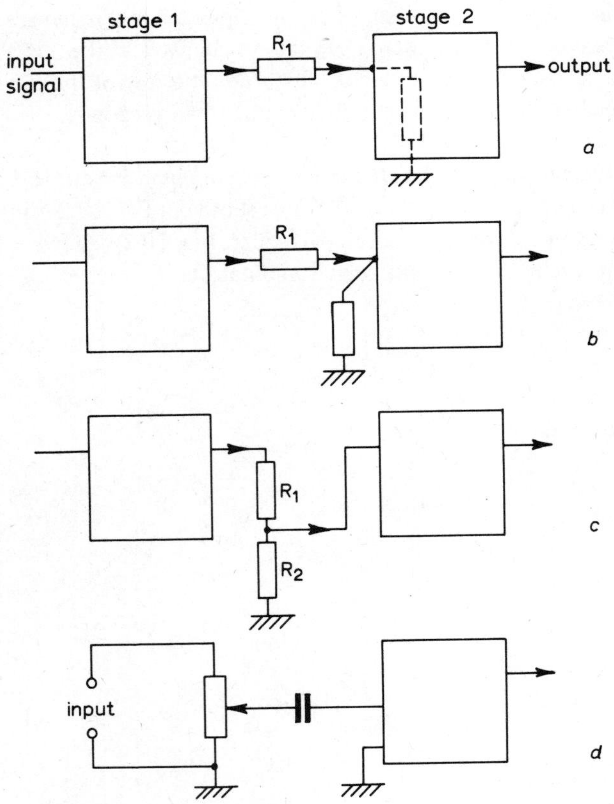

Fig. 10.3 Resistive coupling developed to show attenuator effect

A voltage at the input to R_1 and R_2 will cause a current to flow and the volt drop across R_2 – the output voltage – compared with the whole input voltage is in the same proportion as R_2 is to $R_1 + R_2$.

i.e.

$$\frac{\text{Output voltage}}{\text{Input voltage}} = \frac{R_2}{(R_1 + R_2)}$$

This applies to both a.c. and d.c. if the input should be a mixture as was assumed in Fig. 10.1 *a*, where, had potential-divider coupling been used to reduce the 8 V d.c. level down to 0·5 V d.c. – a ratio of 16:1 – then the alternating voltage would also have been reduced to one-sixteenth of its former value. In an audio amplifier the alternating voltages require amplification from stage to stage, not deliberate attenuation, as a rule.

One important instance of where attenuation (the opposite of amplification) is needed is when an output is too large and a volume control or gain control is used. A potential-divider network is then deliberately included – usually with a coupling capacitor to block d.c. flow – around the beginning of the amplifier chain as shown in Fig. 10.3 *d*.

Using resistor – resistor coupling there is no distortion of wave shape but only reduction of amplitude.

10.5 *RC* COUPLING OF SQUARE WAVE – INTEGRATING

If a capacitor is deliberately added in place of R_2 in Fig. 10.3 *c*, the function of the coupling circuit is changed once again. A capacitor stores electrical charge, and, if fed from the changing voltage levels of a square wave to *C* via a resistor, the capacitor requires time to change its state-of-charge and so to follow the input waveform.

Assuming that the square wave in Fig. 10.4 *b* is applied to the circuit in Fig. 10.4 *a*, a sudden rise of voltage and then an interval while the voltage remains high and steady sees charge accumulating on *C* through *R* until *C* finally manages to reach about full voltage. The sudden drop of input voltage and a period of zero value produces an oppositely shaped discharging curve.

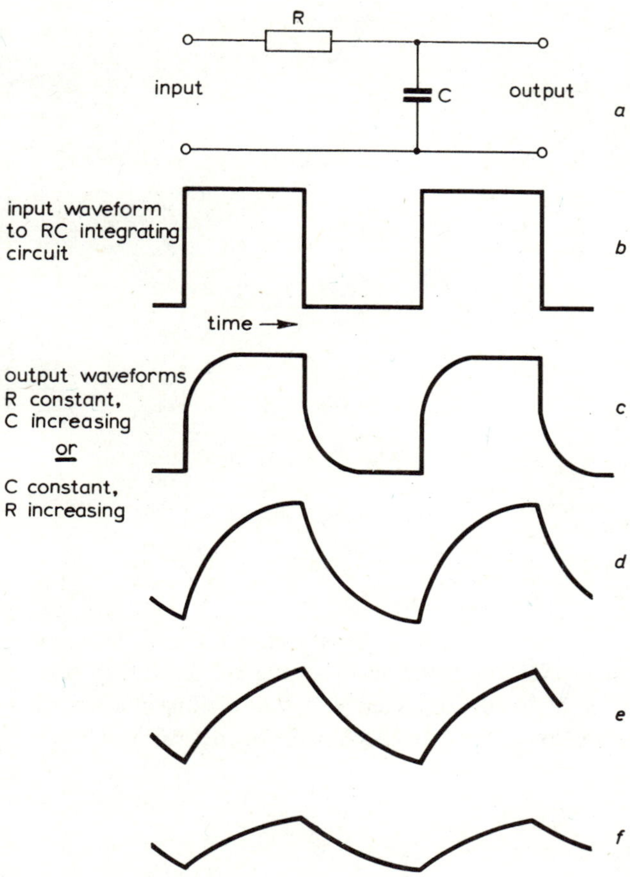

Fig. 10.4 The result of feeding a square wave into an *RC* integrating circuit

Now an integrating *RC* circuit produces the reverse effect of a differentiating *CR* circuit. Where the latter provides an output which is a measure of the rate of change of the input, in the integrating circuit the output is the successive addition of areas under the voltage-time graph. Four curves are shown, Fig. 10.4 *c* to *f*, in which the output waveform varies from an almost undistorted replica of the input waveform to a low amplitude triangular shape, depending upon the *RC* combination in use.

Remember that a sine wave fed to an integrating circuit suffers no distortion but there will be some attenuation and phase shift depending on the *RC* values used.

10.6 DIODE SHAPING CIRCUITS

A diode is similar to a switch that closes and becomes almost a short circuit when the anode (blunt end of arrow)

becomes more positive than the cathode (sharp end of arrow), or if the cathode becomes more negative than the anode, which is another way of saying the same thing.

Fitted across the resistor of the *CR* coupling network in Fig. 10.5 *a* a diode would not affect the output waveform during interval O–P in Fig. 10.5 *b* since the diode would switch itself open, but from P–Q the diode becomes effectively a short circuit and the output waveform stays at zero instead of going negative. The diode thus 'CLIPS off' the negative-going waveform at the output.

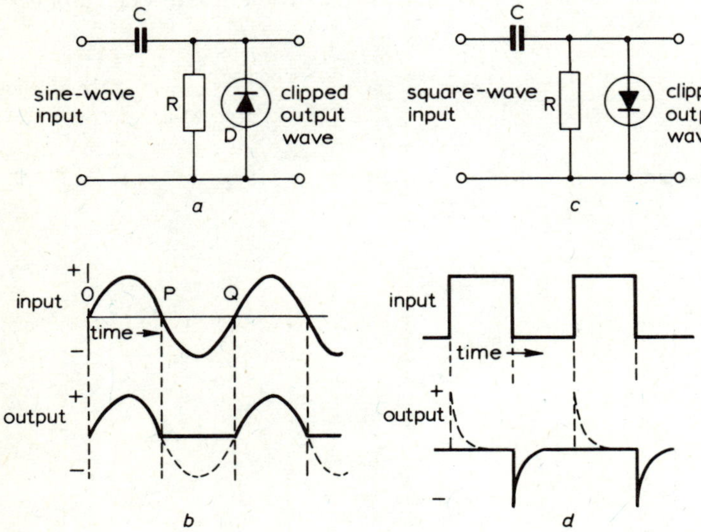

Fig. 10.5 Showing the effect of adding a clipping diode
a and *b* Circuits coupling a sine wave
c and *d* Differentiating a square wave

If a similar coupling circuit in Fig. 10.5 *c* has a pulse or square waveform at the input but with the diode connections reversed, the result this time is that the positive-going waveforms are clipped, leaving the negative spikes only (this is a differentiating circuit) as in Fig. 10.5 *d*. A rather crude, but effective, way of looking at a diode is to consider the positive waveform going through in the direction of the arrowhead while the negatives cannot.

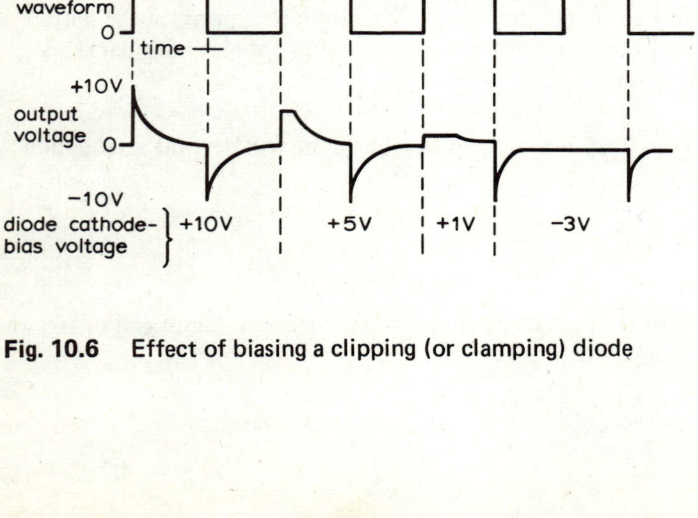

Fig. 10.6 Effect of biasing a clipping (or clamping) diode

Instead of connecting one end of the diode to chassis, this end might be connected to a voltage supply such that the diode's start-of-clipping is determined by the particular voltage level. This is illustrated with the diode in Fig. 10.6 where the output waveform swings between +10 to −10 V and the diode bias voltage may be varied.

At first the diode is connected to +10 V; the anode (sharp end) only just reaches +10 V due to the output pulse and the pulse is thus unaffected by any diode action. As the diode bias voltage is reduced to, say +5 V, the upper portion of the spike is clipped off at the +5 V level; the diode may be said to limit the output at +5 V, leaving the lower portions of the waveform unaffected.

Reducing the bias to zero and then to −3 V, the top or positive part of the waveform is progressively clamped down or clipped off at the value of the bias voltage. Another description of the diode action is LIMITING since the top of the swing or peak-to-peak travel of the waveform is limited in size; clipping, clamping and limiting are all similar in action. Reversing the diode removes the negative peak and leaves the more positive part of the waveform unaltered.

10.7 USES OF DIODE CLIPPING ACTION

A combination of large and small pulses could be separated by a clamp diode circuit so that only the tips of the larger pulses in Fig. 10.7 remain after passing through a suitable coupling and biased-diode circuit.

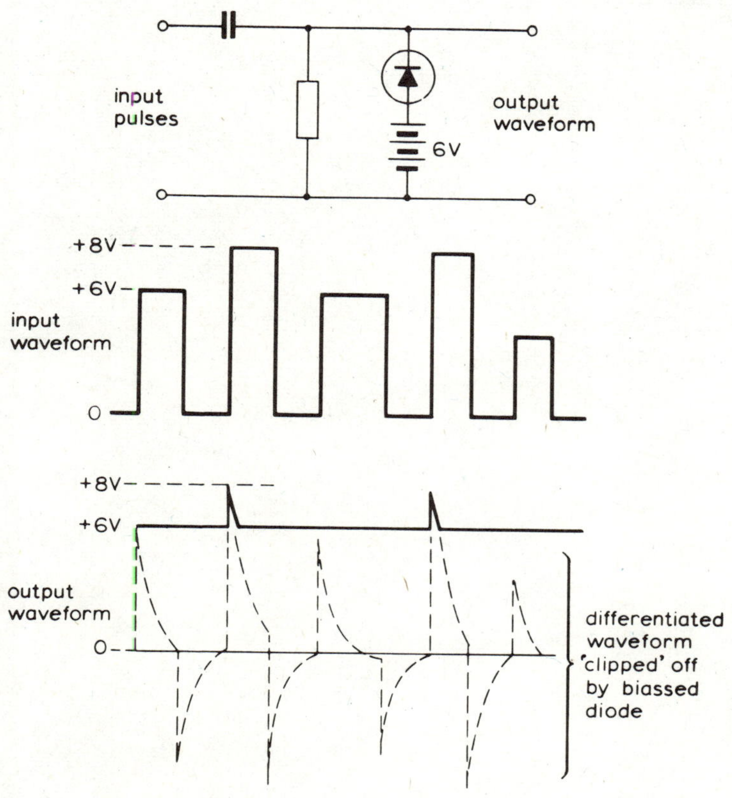

Fig. 10.7 Separation of different amplitude pulses by a suitably biased diode

A simple revcounter could use the circuit of Fig. 10.8; R_1-D1 and the battery form an amplitude limiting circuit so that regardless of the input amplitude, only pulses of 6 V amplitude are applied to the differentiating and negative-spike-clipping arrangement of C_1-R_2-D2. The resulting constant-amplitude constant-width pulses are fed to the indicator via a full-scale adjuster, R_3.

This circuit is also an f.m. demodulator (or frequency discriminator) because variations above or below an average frequency produce a proportional increase or decrease of current through the load; variations of input pulse amplitude have no effect because of D1's action in clipping all inputs to the same amplitude.

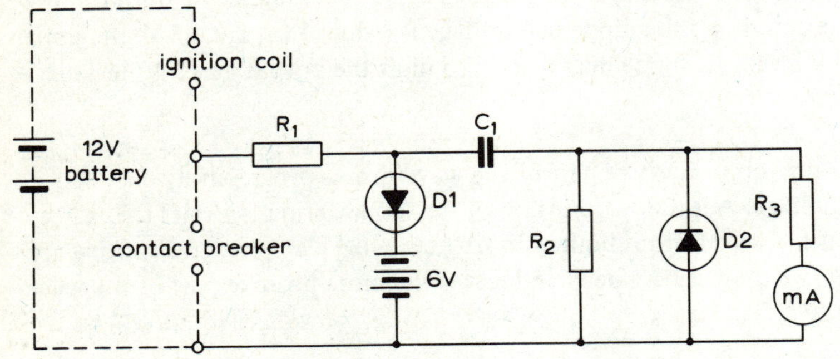

Fig. 10.8 Simple circuit of revcounter or tachometer

In an oscilloscope (Chapter 14) some of the Y-input waveform is grossly overamplified so that the result is of virtually square-topped pulses. These pulses may be differentiated into spikes, one polarity 'lost' by the use of a clipping diode and the remaining spikes used to keep the timebase in synchronism with the Y-input waveform.

Chapter 11

Power supplies

11.1 INTRODUCTION

First, make sure that the difference between a.c. and d.c. is remembered:– alternating current, produced by an alternating voltage, repeatedly changes direction or polarity and the waveform is seen to go above and below zero, as in Fig. 11.1 *a*. Direct current produced by a direct voltage flows in one direction only and the waveform is seen to remain on one side only of the zero voltage axis, as in Fig. 11.1 *b*. In neither instance, a.c. or d.c., is there any mentio of amplitude or actual waveform and this is the reason for showing 'impure' waveforms in Fig. 11.1; in particular, d.c. could go down to zero, temporarily switched off as in Fig. 11.1 *b*, but as long as the flow is always in one direction or of the same polarity, then the result is d.c.

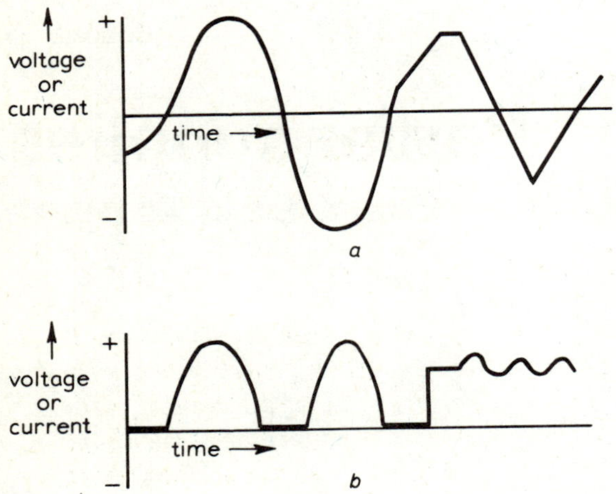

Fig. 11.1 Waveforms emphasising the difference between alternating and direct values

An impure d.c. waveform could be looked at as pure d.c. with a.c. superimposed as shown in Fig. 11.2; pure d.c. is not directly obtained from a.c. supplies and it is often useful to be able to think of the two being combined in the impure d.c. state but then losing the a.c. part to leave pure d.c.

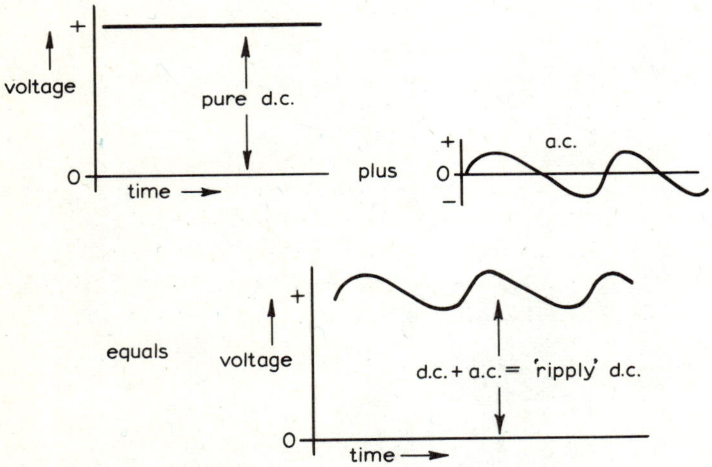

Fig. 11.2 Combination of a.c. and d.c. to give 'hummy d.c.'

11.2 RECTIFIER DIODES

These have the symbol shown in Fig. 11.3 and may be regarded as automatic switches, closing for electron flow *against* the 'arrow' and opening when the flow tries to go *with* the 'arrow'.

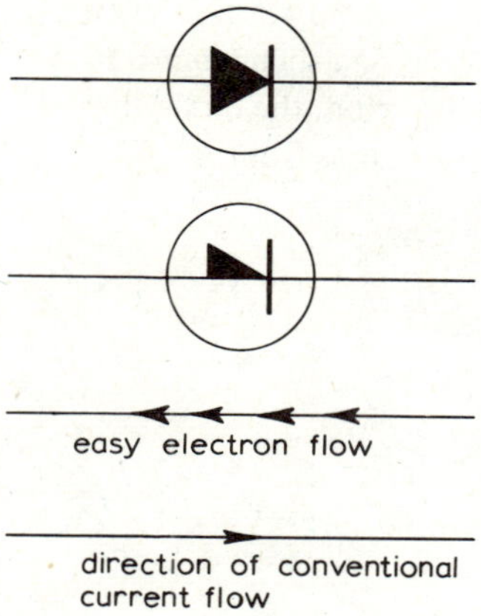

Fig. 11.3 Diode symbols and direction of 'easy flow'

11.3 THE TRANSFORMER

The transformer symbol, Fig. 11.4 is really self-explanatory; two coils are wound on a laminated iron core, the PRIMARY coil having the applied voltage – usually the mains supply when considering power supplies – connected to it while the SECONDARY winding provides the output voltage. The number of turns on primary and secondary coils determines the voltage transformation ratio, a STEPDOWN ratio of 10:1 giving a secondary voltage of 24 V a.c. from an input or primary voltage of 240 V a.c. The advantage of using a transformer is that the only coupling or connection between the mains and the secondary coil is the varying magnetic field within the iron core. There is no electrical connection whatever between primary and secondary windings and so no danger of a mains-supply shock to the person using the secondary supply.

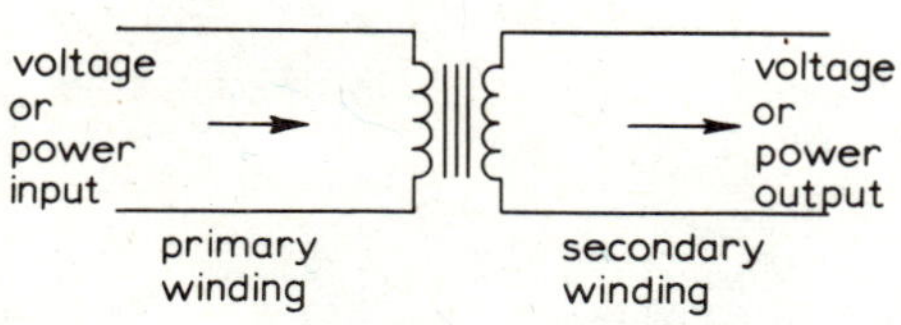

Fig. 11.4 Transformer symbol

11.4 RECTIFICATION

Fig. 11.5 *a* gives the basic block diagram of a rectifier circuit but a transformer will always be included in other diagrams, as Fig. 11.5 *b* for safety and isolation purposes. Although most t.v. receivers derive their internal d.c. supplies directly from the mains a.c. supply it is essential for safe t.v. servicing to include an isolating transformer of 1:1 ratio, hence its inclusion in all diagrams.

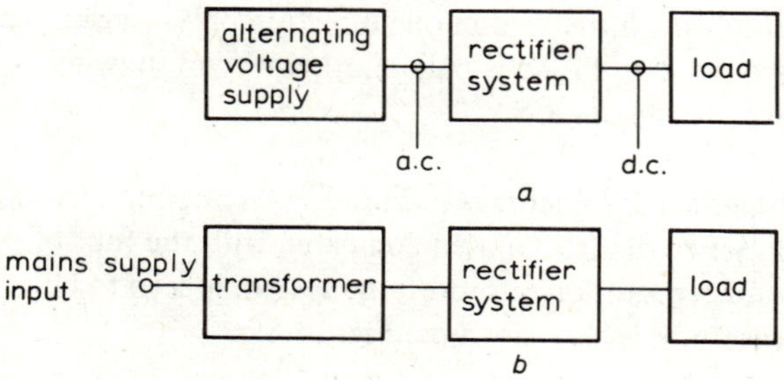

Fig. 11.5 Power supply block diagram

The actual circuit of a simple rectifier system is given in Fig. 11.6*a*. The output from secondary A-B is an alternating voltage so that current will try to flow from A to B for one polarity of half cycles and then from B to A for the other half cycles. The rectifier diode only allows electron flow in the B-to-A direction, the diode then switching itself open when the voltage at A-B reverses and the electron flow would try to pass from A-to-B.

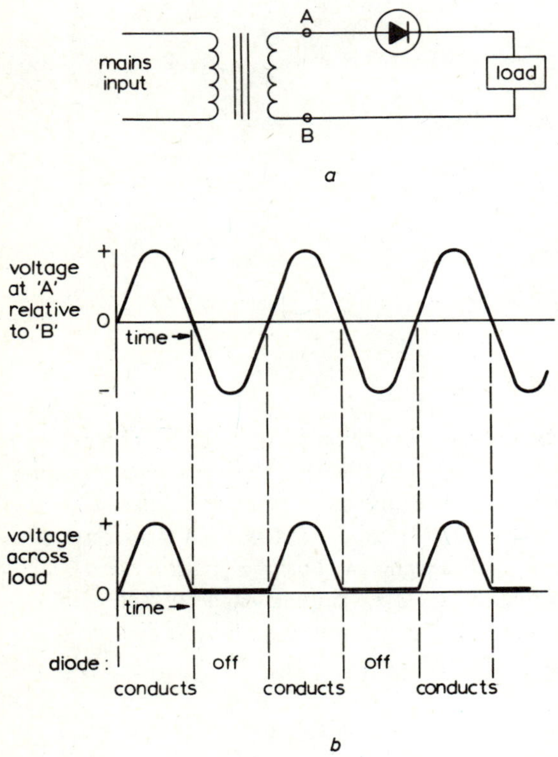

Fig. 11.6 *a* Half wave rectifier circuit
b waveforms

The current through the load is thus a series of unidirectional pulses, according to the definition of d.c. at the beginning of this chapter. Between the pulses of current there is a gap where the diode allows no flow. The voltage waveform across the load is compared with the voltage A-B in Fig. 11.6*b*; this system is called a HALF-WAVE RECTIFIER system, the title indicating the waveform.

The half-wave system is the simplest but has the disadvantage that for half the time there is no current flow through the load. The variation of the d.c. level is called RIPPLE (or HUM).

Using the same transformer secondary voltage a four-rectifier circuit may be added to give FULL-WAVE rectification as shown in Fig. 11.7*a*. The rectifier diodes are in a 'bridge' or diamond formation and their automatic opening and closing results in the waveform of Fig. 11.7*c* being developed across the load. Every half cycle of the input is made use of, the diodes steering the current pulses through the load in one direction only – again d.c. derived from a.c. – although there is still a lot of ripple on the d.c. Fig. 11.7*b* is only another way of drawing the 'bridge' and is called a 'lattice' formation; the action is exactly the same.

There is another full-wave rectifier system in use called the biphase (or sometimes push-pull) system, but a transformer with a centre-tapped secondary is needed although only two diodes are used compared with the four of the bridge rectifier system. The circuit is given in Fig. 11.8*a* for recognition purposes; the system is seen to be the same as in Fig. 11.7, and the full-wave rectifier output waveform is also the same, Fig. 11.8*b*.

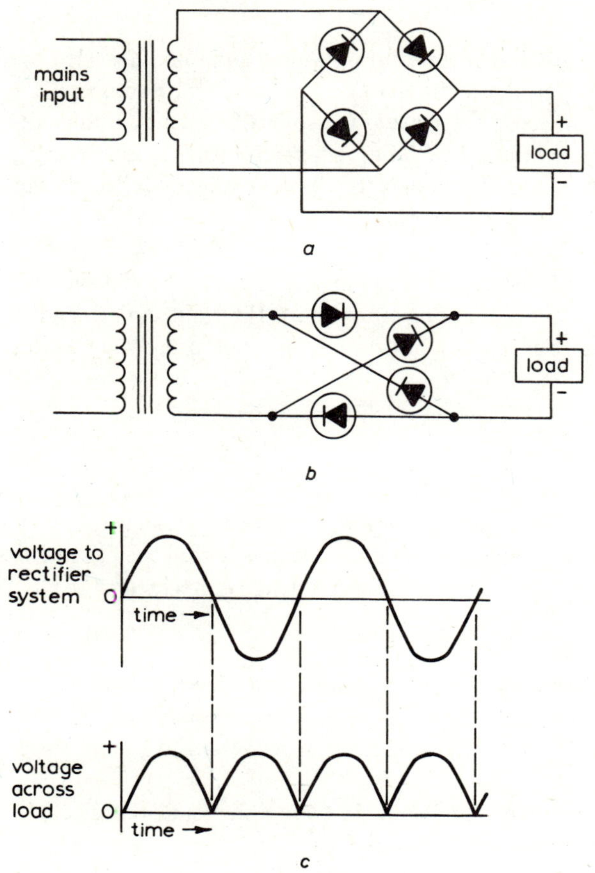

Fig. 11.7 Full wave rectifier circuits and waveforms

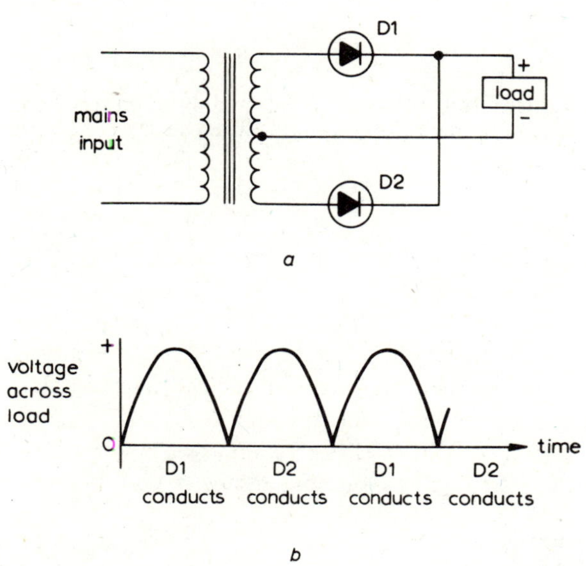

Fig. 11.8 Full wave rectifier circuit using a biphase system

11.5 RIPPLE FREQUENCY AND SMOOTHING

Although the output from a rectifier system is d.c. it is not sufficiently smooth to supply equipment designed to operate from a 'pure d.c.' supply such as a battery. The ripple on the rectified d.c. is excessive and results in a loud hum in radio or amplifier loudspeakers. The ripple frequency is the same as the supply frequency if the half-wave system is used – there is one rectified cycle per a.c. cycle in Fig. 11.6 *b* but, when a fullwave system is used as shown in Fig. 11.7 *b*, there are two ripple cycles per single a.c. cycle, so that the ripple frequency is double the supply frequency.

To convert the 'ripply' d.c. to pure d.c. a SMOOTHING circuit is added between the rectifier system and the load. Considering the hummy d.c. to comprise [pure d.c. plus a.c.] the smoothing circuit must filter out the a.c. and leave the d.c. for the load. It is difficult to filter out absolutely all the a.c. but the smoothing is usually adequate to reduce the ripple to a level where it is not objectionable.

11.6 REGULATION

If a simple cell has the voltage measured off-load and again when supplying a load the on-load voltage is found to be the lesser value. The voltage reduction on load is caused by voltage drop in the resistance within the cell and is a common cause of distortion in pocket radios. The battery voltage must be measured while connected to the radio and supplying current.

A rectifier system also includes resistance within the transformer, rectifier and smoothing systems, and the voltage on-load will be lower than the off-load output voltage. The REGULATION of the power supply system is taken as the voltage reduction on-load compared with the off-load voltage and is often converted to a percentage.

Consider a system as shown in Fig. 11.9 *a* where the off-load voltage is 25 V d.c. but when the load is connected the voltage drops to 22 V d.c. The full-load regulation of the complete system is then

$$\frac{25-22}{25} = \frac{3}{25} = 0{\cdot}12$$

or, as a percentage

$$\frac{3}{25} \times 100\% = 12\%$$

Using the simple transformer – rectifier – smoothing system, the output voltage depends mainly upon the transformer windings. To reduce the voltage across the load to lower than the normal full value might be achieved by

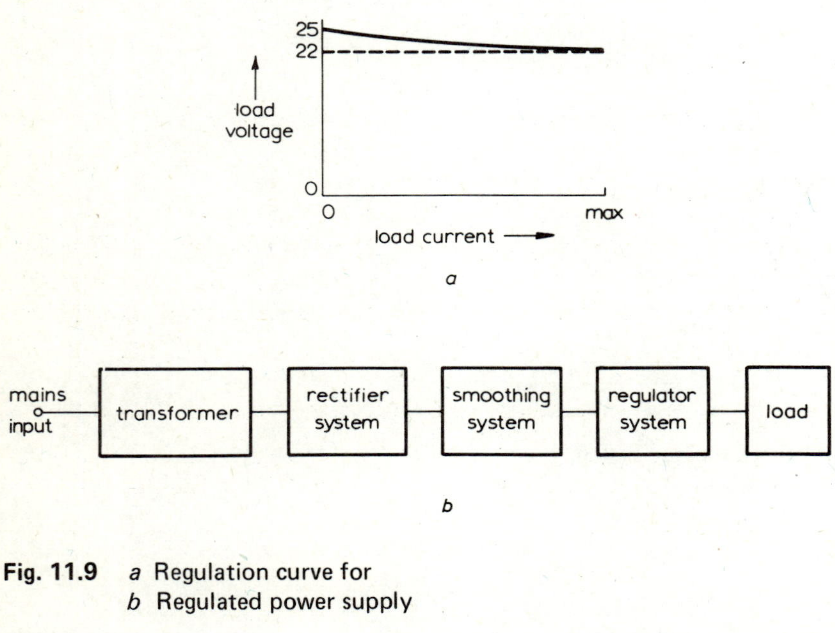

Fig. 11.9 *a* Regulation curve for
b Regulated power supply

including a resistor of suitable value as the regulator block in Fig. 11.9 *b*, but frequently a valve or transistor is used since the regulation is more easily varied and controlled.

11.7 POTENTIAL DIVIDER

The simplest method of obtaining a low voltage from a higher voltage system is to use a simple series resistor as shown in Fig. 11.10 *a* and providing that the load current is known and constant this system works well. However, if the load varies, the drop in voltage across R also varies and so the load voltage varies too. The next simplest method to obtain a lower voltage, and one that achieves better regulation is to use two resistors as in Fig. 11.10 *b*. The current bypassed or 'bled' through R_2 partially compensates for variations of load current and thus results in better regulation.

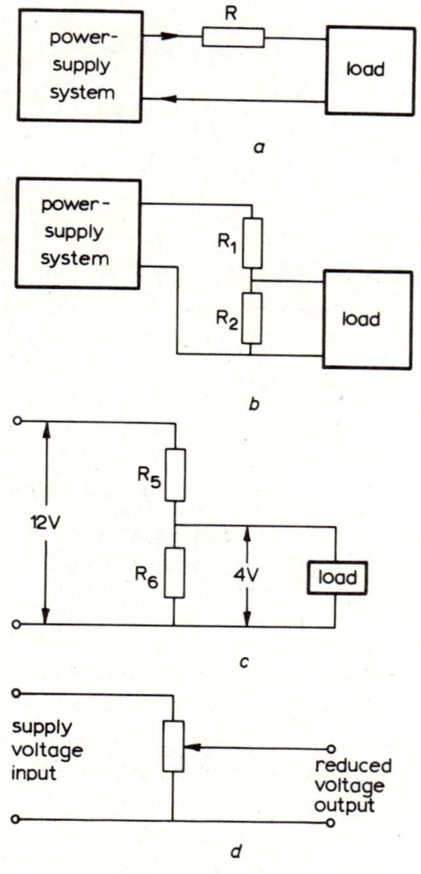

Fig. 11.10 Methods of reducing output voltage

In Fig. 11.10 *c* a 12 V supply is to provide a negligible current load at, say 4 V. Now 4 V is one third of 12 V and so R_6 should be one third of the total resistance $(R_5 + R_6)$. If R_6 is say, 1000 Ω then $(R_5 + R_6)$ must be 3000 Ω, thus R_5 is 2000 Ω. These two resistors of 1000 and 2000 Ω will form the potential divider circuit to give the wanted 4 V between the junction of R_5/R_6 and the bottom of R_6.

If 8 V were needed, it is now seen that an interchange of resistors to use 1000 Ω at the top and 2000 Ω at the bottom of the R_5/R_6 system would then give

$$\frac{2000}{2000 + 1000} \times 12\,\text{V} = 8\,\text{V}$$

across the lower resistor.

Note that the load current is to be small compared with the bleed current; or put in another way, the load resistor must be considerably larger than the lower resistor R_6 for the simple calculations to be correct. Also, the calculated resistors rarely turn out to be 'preferred values', and it may be better to use a wirewound potentiometer as both R_5 and R_6 together, with the sliding contact being set to give the required voltage. This system is shown in Fig. 11.10*d*.

11.8 STABILISATION

If a power supply is built and found to be of poor regulation it is possible to add feedback so that the output voltage is increased when the loading tends to make the voltage droop and similarly the output voltage is held down as the loading decreases. The block diagram of this system is shown in Fig. 11.11; the comparator block compares the actual output voltage with a voltage derived from a standard and passes a correction voltage signal to the regulator block.

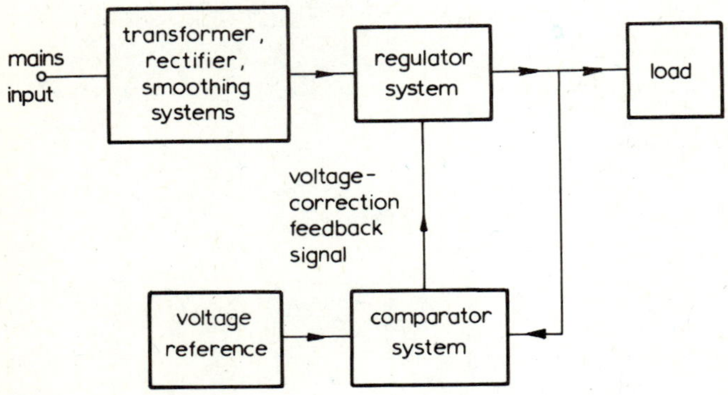

Fig. 11.11 Regulated and stabilised power supply

Using a stabilised and regulated supply the regulation of a poor system may be improved to 0·01% against variations in either mains supply or load variations. Of course, the better the regulation and stabilisation the more expensive the power supply becomes.

11.9 SAFETY

The full circuit of a power supply must contain switching, fuses and pilot lamps for initial safety. Inside the power supply there will probably be capacitors which are used for charge storage, and so, may remain charged for weeks after switchoff, unless fitted with discharge or 'bleeder' resistors of suitable value.

If you suspect a power supply to be faulty, check the output voltage with a voltmeter (use a suitable range of course, a.c. or d.c. as needed) and see that the pilot lamp is not open circuit. A dark pilot lamp need not mean a 'dead' power supply, so switch off and isolate the input supply (pull the plug out if there is one) before checking the fuses and lamp for continuity with an ohmmeter. If you have to remove covers or panels then use one-handed working inside where possible; short circuit the capacitors with prods connected by well-insulated wire. Never replace fuses with a higher value rating so that you 'hide' the fault, since this could lead to serious overloading and damage. Always maintain the pilot or warning lamps in good condition. They are the main indicators of good/bad conditions within apparatus.

Chapter 12

The radio receiver

12.1 INTRODUCTION

Information may be transmitted through space by radiating r.f. from an aerial. The aerial is energised by an amplifier fed from an r.f. oscillator and the information is modulated on to the carrier usually as either amplitude or frequency modulation, a.m. or f.m.

At the receiving end, a very small voltage – probably only one or two microvolts – is induced into the receiving aerial system and although the voltage is an exact replica of the transmitted modulated carrier there are certain to be several – or many – other carriers on the same waveband also being received, so a frequency selection process is needed as in Fig. 12.1.

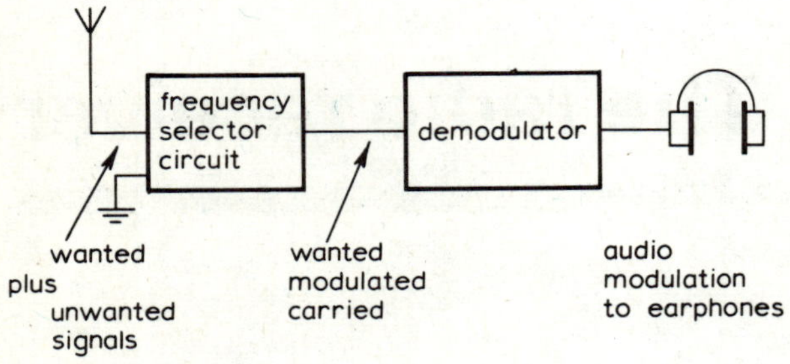

Fig. 12.1 Basic receiver block diagram

12.2 STATION SELECTION

A selective tuned circuit consists of a coil (or inductor) and a capacitor, usually connected in parallel as shown in Fig. 12.2*a*. Either component may be variable to adjust the tuning to a particular frequency; up to ten years ago the capacitor was the easier component to vary but now the coil often has a moveable core to vary the inductance, particularly in car radio receivers. An arrow through the capacitor symbol shows that it is variable as in Fig. 12.2*b*, while in Fig. 12.2*c* the iron dust or ferrite core is shown as being of variable position.

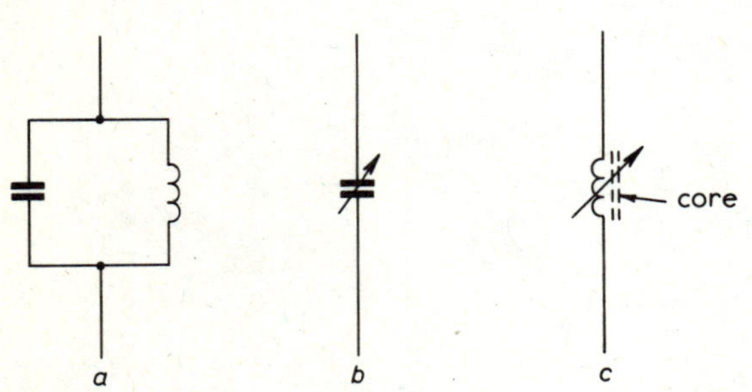

Fig. 12.2 Fixed tuned circuit together with 'variable' symbols

The simple 'crystal set' has already been described in Chapter 7 and is a basic circuit of Fig. 12.1, but there is the disadvantage that no amplification is used before demodulation (or detection) and so the very small voltages available from the aerial can barely drive the demodulator stage. If the circuit described were built you would probably receive only one or two stations and the stronger one would be hearable over much of the tuning range. To be able to receive more stations the receiving apparatus needs SENSITIVITY and then to be able to sort out one station from a neighbour requires SELECTIVITY otherwise all the signals will be heard together in the loudspeaker.

Sensitivity may be obtained by amplifying the aerial voltages up to a higher level, using one or two valve or transistor voltage amplifiers. It is no good however, amplifying every signal that arrives via the aerial, and tuned circuits of Fig. 12.2 are included to select and reselect the wanted frequency while at the same time repeatedly rejecting the unwanted frequencies. With only two amplifying stages a voltage gain of 500 to 1000 is obtainable and 1 μV at the aerial becomes 1 mV which is able to drive the demodulator stage more successfully. Three tuning circuits are used in Fig. 12.3 and now one of the difficulties in the design of a good TUNED RADIO FREQUENCY (t.r.f.) receiver is seen; the three circuits must be kept exactly in tune with each other as the tuning dial is rotated from one end of the waveband to the other.

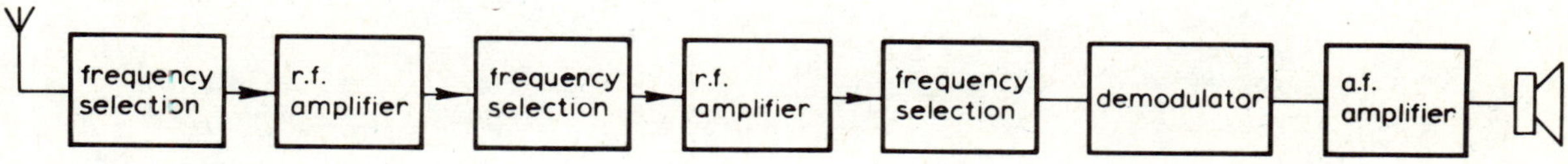

Fig. 12.3 Multistage 'tuned radio frequency' or t.r.f. receiver

12.3 DISADVANTAGE OF THE T.R.F. SYSTEM

Three – and four – gang tuning capacitors are made and are quite cheap, but in the t.r.f. the tuned circuit – amplifier – tuned circuit physical layout would need to be absolutely symmetrically placed on the chassis for all wiring to be exactly the same length, and this is very difficult to arrange. Early t.r.f. receivers used a separate tuning control for each circuit – sometimes four dials in a row along a polished ebonite panel 50 cm long – and each could be adjusted for best results. Obviously not the type of control panel for a housewife or a lorry driver who wishes to change stations in a hurry!

A further disadvantage is the possibility in a compact circuit of feedback from the output back to the aerial input stage. Either type of feedback, negative or positive, is a nuisance and does not help the performance of the receiver, but, in particular, positive feedback creates oscillations at the frequency to which the receiver is tuned and the receiver then becomes a low power transmitter and most unwelcome to neighbouring listeners.

The t.r.f. receiver is simple to build but to gain sensitivity and selectivity requires several stages of amplification and tuning, respectively. These bring problems of stability and ganging and about 1928 the supersonic heterodyne or SUPERHET receiver was designed, and being much superior to the t.r.f. is now used in nearly 100% of receiver designs.

Chapter 13

The superheterodyne receiver

13.1 HETERODYNES

Listening to a radio at the 1500 kHz (or 200 m) end of the m.w. band many whistles are heard, caused by the many station carriers present and within a few kilohertz of each other. Suppose that one carrier of 1450 kHz and a second carrier of 1452 kHz arrive at the demodulator stage – a nonlinear stage, not just an amplifier – together, then a beat-note, heterodyne or whistle of 2 kHz is heard from the loudspeaker as well as the modulation derived from the individual signals. These audio heterodynes are part of the modulation, demodulation or mixing process described in Chapter 6 and may be further demonstrated by listening to a station with a radio while tuning an r.f. generator to nearly the same frequency as the station's carrier. The resulting audible heterodyne depends upon the difference in frequency between the two carriers received by the radio.

There is an inaudible heterodyne generated if the frequency is greater than about 15 kHz and so 'above audio' or 'supersonic'. Thus tuning the radio receiver to 1400 kHz and setting the nearby r.f. generator to 1850 kHz will result in a heterodyne of 1850 −1400 = 450 kHz being generated, even though there is no apparent difference in the operation of the receiver to which you are listening.

13.2 SUPERHETERODYNE RECEIVER

A fixed-frequency, high-gain and selective r.f. amplifier with ample screening to prevent feedback is fairly easy to construct. The wanted station's frequency is converted or changed to that handled by the fixed-frequency or INTERMEDIATE FREQUENCY amplifier and after undergoing gain and selection from this amplifier the signal is then demodulated and fed to the loudspeaker via an audio amplifier. A simple block diagram is shown in Fig. 13.1.

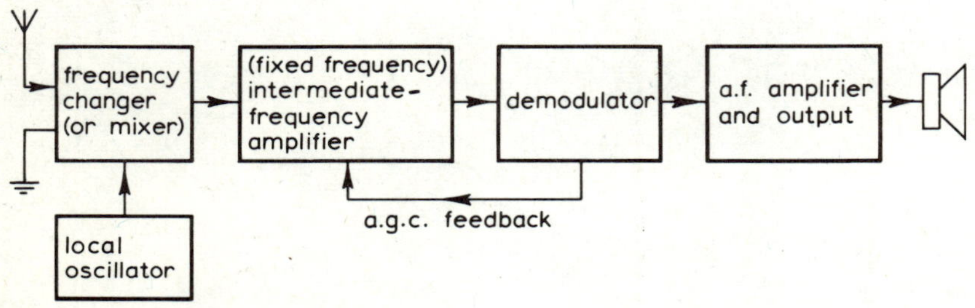

Fig. 13.1 Basic superhet receiver

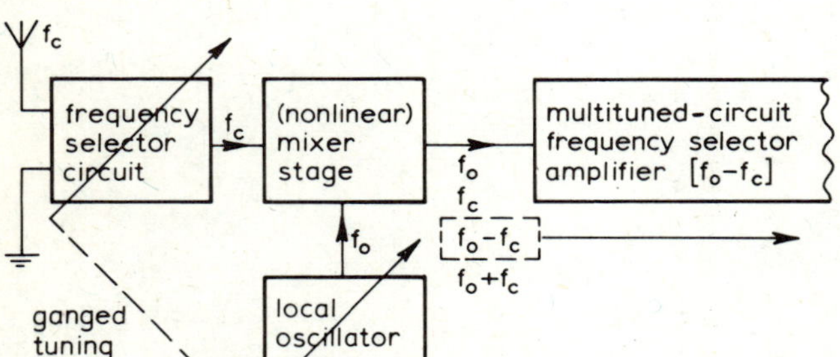

a

received wavelength (metres)	received frequency f_c (kHz)	local oscillator f_o (kHz)	intermediate frequency $f_o - f_c$ (kHz)
1500	200	670	470
600	500	970	470
300	1000	1470	470
200	1500	1970	470

b

Fig. 13.2 Relation between frequencies at the aerial, local oscillator and mixer stage output

The MIXER or 'first converter' takes the incoming signal f_c and a locally generated frequency f_o, mixes them together in a nonlinear stage (as described in Chapter 6) and produces at the output the main four products as shown in Fig. 13.2*a*.

The local oscillator frequency, f_o, is higher than the incoming carrier f_c, and f_o and f_c are calculated such that the difference frequency, $f_o - f_c$, is the same as the intermediate frequency (or i.f.). A list of received station wavelengths and frequencies together with the required local oscillator frequencies to produce a typical i.f. of 470 kHz is shown in Fig. 13.2*b*. Still a very small voltage, the changed-frequency signal is now amplified, filtered and selected within the i.f. amplifier by at least three or four fixed-tuned circuits and amplifiers which are able to remove other signals that may be near but not exactly on the same frequency as the wanted signal.

The demodulator stage also produces a direct voltage that is proportional to the strength of the signal and this d.c. is fed back to the i.f. amplifier stages in Fig. 13.1 to reduce the gain with strong signal inputs but allow the gain to be maximum for weak signals; a.g.c. was described in Chapter 9.

After demodulation the audio signal is amplified and fed to a power amplifier and thence to a loudspeaker, a diagram indicating the transistors and tuned circuits used being given in Fig. 13.3.

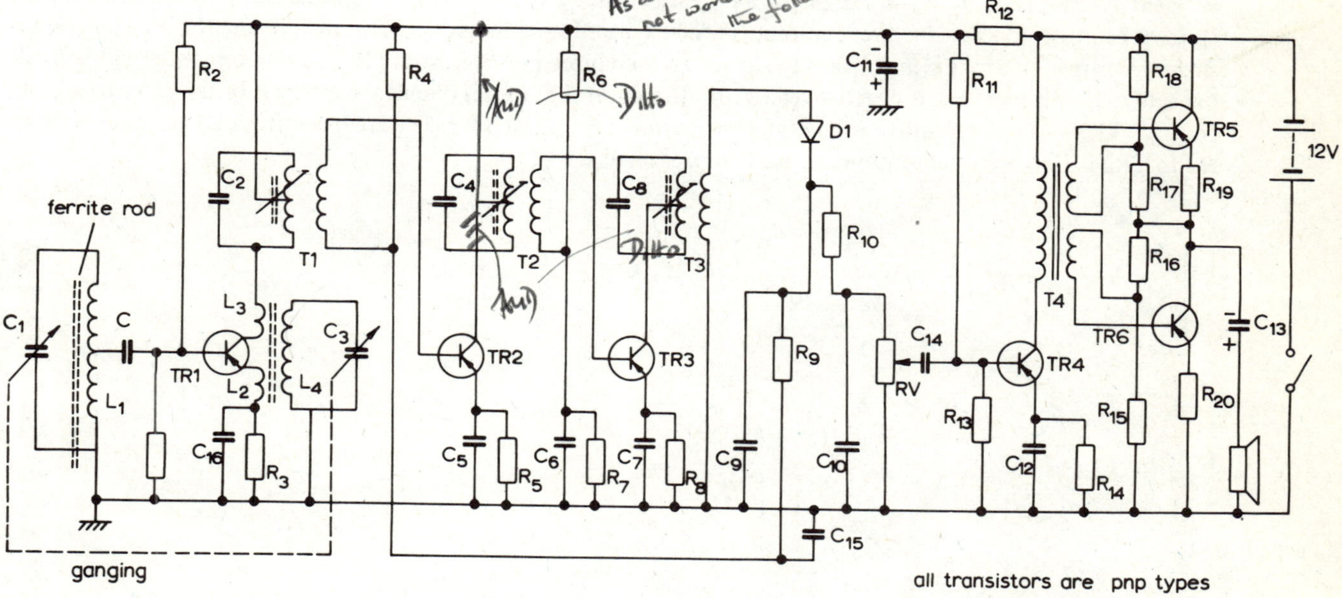

Fig. 13.3 Circuit of simple superhet receiver

Apart from including an f.m. demodulator stage instead of an a.m. type, a v.h.f. f.m. receiver is very similar to a medium/long wave a.m. receiver. The v.h.f. receiver still uses a.g.c. to equalise the output on strong and weak stations but may also have an automatic frequency control circuit added from the demodulator stage back to the oscillator to minimise the effects of oscillator drift.

Because of the lack of interference on v.h.f. and the relative lack of station congestion, higher quality audio is available from the wideband f.m. transmissions. Many radio receivers are now able to receive both a.m. and f.m. programmes (though not at the same time, of course!) and it is common to include a push-pull output stage for both operating-efficiency and the ability to provide high-quality, low-distortion output.

13.3 DESCRIPTION OF FIG. 13.3 CIRCUIT

This figure is of a much simplified superhet receiver. No long/medium switching is shown, this involves slight modifications only to L_1 and L_4 to change selected frequencies, as appropriate.

TR1 is doing two jobs. Firstly, oscillating because there is positive feedback from L_3 to L_4 (tuned) to L_2, with power gain of TR1 allowing the frequency selected by L_4–C_3 to maintain continuous oscillations. Secondly, the frequency selected by L_1–C_1 is fed to the base of TR1 and there 'mixed' with the oscillation frequency to provide at the collector the various frequencies shown in Fig. 13.2*a*.

T1 primary is tuned by C_2 and the particular intermediate frequency for this receiver of 470 kHz (plus sidebands) is selected and passed on to TR2. The small i.f. voltage at the base of TR2 is amplified by TR2 and reselected by *T2*–C_4, rejecting adjacent frequencies that belong to unwanted transmissions that were also frequency-changed in TR1. A rather larger i.f. voltage is then presented to the TR3 base, amplified and again selected by T3–C_8.

The modulated carrier of the wanted station is now demodulated by D1, 'rectified r.f.' flowing through R_{10}–R_V, the r.f. pulses being largely 'smoothed out' by C_9–C_{10} so that across the volume control R_V there is an audio waveform, the original modulation.

TR4 is an audio amplifier; T4 is a phase-splitter transformer producing two equal (but opposite phase) signals to feed the push-pull output stage, TR5–TR6. The loudspeaker in this instance is capacitively coupled to the collector-emitter junction of TR5–TR6, a 'transformerless' output stage.

Whereas TR3 is biased by R_6–R_7 acting as a potential divider, TR2 is biased by R_4 and R_9–R_{10}–R_V. As a received signal is being demodulated a direct current is produced to flow through R_{10}–R_V, making the R_9–R_{10} junction slightly positive. A strong signal makes this junction even more positive. Now TR2 is a pnp transistor and normally expects to have a base that is negative relative to its emitter, but in this circuit the strength of the received signal controls TR2's negative/positive voltage and so controls the gain of TR2. The feedback from D1–R_{10} junction via R_9 to TR2 is thus seen to be automatic gain control feedback.

Chapter 14

Oscilloscopes

14.1 INTRODUCTION

The oscilloscope (or scope, as it is more usually called) is a most useful piece of equipment, invaluable in tracing the progress of a signal waveform on its way through a system and showing the results of a fault or bad adjustment within the various stages. However, before dealing with the scope, which demands reasonable skill in adjustment and use, the eyesight of the user must first be considered.

14.2 THE EYE

When you look at anything the rays of light are focussed by the eye onto the retina at the back of the eyeball and nerves here carry the information to the brain. Now when the eye is presented with a very brief flash of information your 'vision system' continues translating the information for about one-fifteenth of a second – this is called PERSISTENCE OF VISION – and at the cinema this situation is made use of when twenty to twenty-five pictures per second are shown on the screen. The brain *thinks* that a continuously changing, moving picture is being shown when in fact a series of still pictures, each slightly different from the previous one, is being seen.

14.3 THE OSCILLOSCOPE

On the face or screen of the cathode ray tube is a single spot of light. This spot of light may be moved around the screen and to the eye and brain it appears that a line is being traced if the spot is moving fast enough and keeps retracing its path over and over again. Presenting a steady trace becomes a matter of controlling the amplifiers within the scope that drive the spot around the screen.

In Fig. 14.1 is a block diagram of the complete scope. The cathode ray tube (or c.r.t.) produces a spot of light that is moved in the vertical direction by the voltage under investigation while the spot is also moved horizontally at the same time by the TIMEBASE oscillator within the scope.

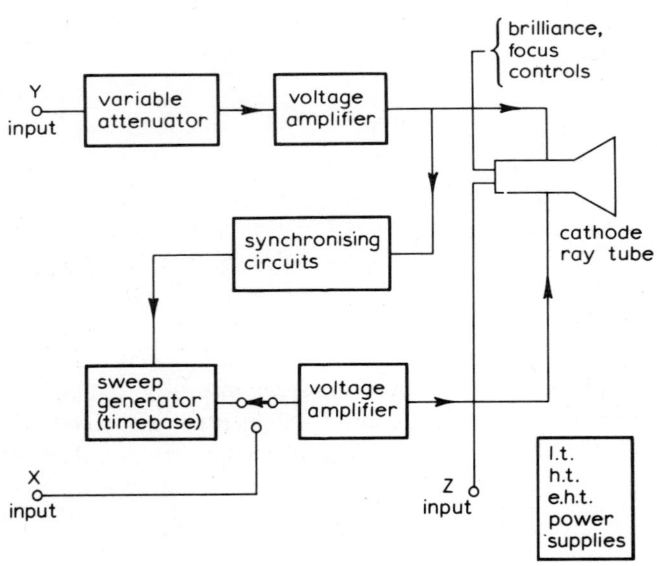

Fig. 14.1 Basic oscilloscope block diagram

Graphs always use the same letters for direction; Y for vertical and X for horizontal, and these letters are also used for scope display directions so that the input voltage to the Y system moves the spot vertically while the time-base is often called the X or SWEEP GENERATOR. A three-dimension graph would also use the letter Z and in a scope this refers to the beam of electrons shot down the c.r.t. to hit the screen and produce the bright spot of light. The Z input or Z modulation, if it is used, refers to the brilliance and could be used to produce a 'dotted' or variable-intensity curve. these directions are shown in Fig. 14.2 *a*, a 'normal' X–Y waveform in Fig. 14.2 *b* and a brilliance or Z modulated version of the same waveform in Fig. 14.2 *c*.

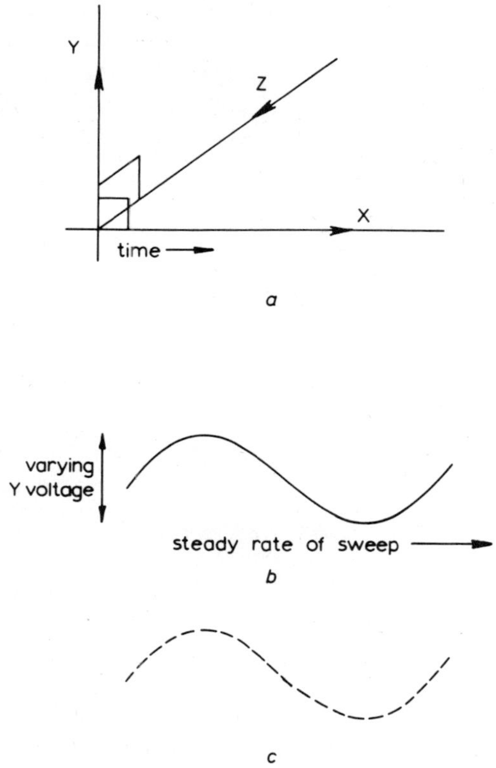

Fig. 14.2 Relation between X, Y and Z axes of c.r.t. screen
Waveform (*b*) uses X and Y while (*c*) is produced using X, Y and Z modulation

14.4 THE ATTENUATOR

The Y amplifier – valve or transistor – is a voltage amplifier with a fixed gain. Usually only about 10 mV is needed at the amplifier's input to produce enough voltage change at the output to move the spot on the c.r.t. by 1 cm., 100 mV being sufficient to move the spot from top to bottom of the c.r.t. face. The amplifier is designed to be as sensitive as this so that small voltages may be investigated easily – but equally well, large input voltage changes will overload the amplifier and be grossly distorted at the output as well as overlapping the c.r.t. screen edges. An attenuator (or voltage divider) is therefore included *before* the amplifier to reduce any large signal voltages down to a set maximum value and this value may then be amplified with minimum distortion and displayed on the screen. The attenuator is usually calibrated – the Y sensitivity control – in volts per centimetre (or volts/cm) so that the centimetre rulings inscribed on the c.r.t. face or transparent screen together with the setting of the attenuator allow swift calculation of the input voltage amplitude.

14.5 MEASURING VOLTAGES

Fig. 14.3 shows an oscillograph – the graph or picture drawn by an oscilloscope – together with the Y attenuator or sensitivity setting. The waveform at the left is 4 cm from peak to peak and so is 40 V peak-to-peak. At the right, the waveform is only about 3 cm from peak to peak and so is of 30 V peak-to-peak amplitude. Remember that the peak-to-peak waveform measured with the aid of a scope will NOT be the same as the voltage measured by a meter due to corrections needed for waveforms and the functioning of the meter.

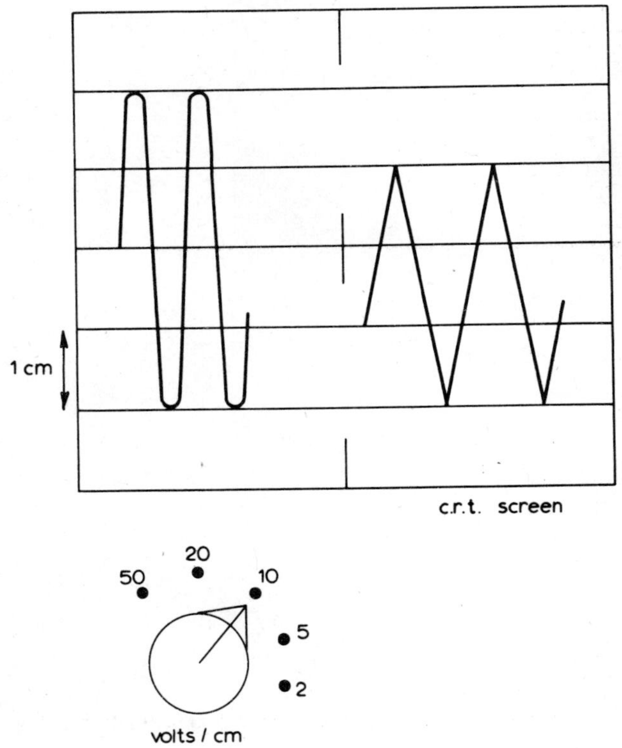

Fig. 14.3 Illustration of voltage measurement

14.6 THE TIMEBASE

The timebase or sweep generator is an oscillator producing a sawtooth or ramp voltage waveform. The voltage increases slowly as in Fig. 14.4 and this slow change moves the spot on the c.r.t. screen from left to right. Note here that a scope is needed to design and view the waveforms present in a new scope model. Ideally the spot should return from right to left in zero time – the flyback – but even electrons take a little time to move so that the flyback part of the waveform is a slope rather than a true vertical.

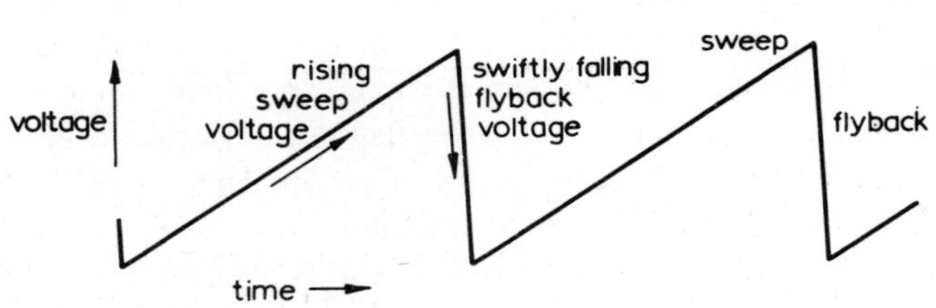

Fig. 14.4 Timebase waveform

The amplitude of the sawtooth generated is usually too small to be used directly and the sweep generator is followed by the X amplifier. The gain of this amplifier is often enough to overscan the c.r.t. and may be variable to allow a waveform being viewed to be adjusted to fit the screen rulings exactly for comparison or measurement purposes.

14.7 MEASURING TIME AND FREQUENCY

The speed of the spot travel across the screen is often set up by a control calibrated in s/cm, ms/cm or μs/cm. In Fig. 14.5 one complete cycle occupies 2 cm, and so 1 cycle lasts for 10 ms. To calculate the frequency, the

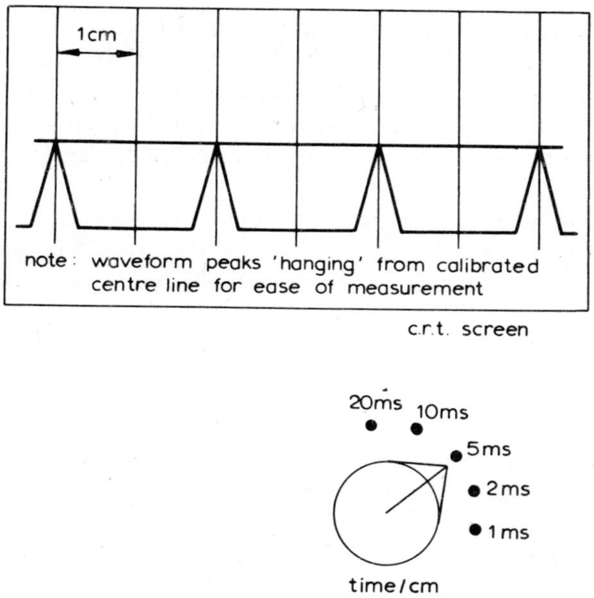

Fig. 14.5 Illustration of time and frequency measurement

measurement of Hertz requires you to know the number of cycles-per-time while the scope display tells you the time-per-cycle.

To convert from the measured

$$\text{time-per-cycle} = 10\,\text{ms}$$

$$\text{i.e. } \frac{10}{1000}\,\text{s/cycle}$$

invert these figures to give

$$\text{frequency} = \frac{1000}{10}\,\text{c/s}$$

$$= \underline{100\,\text{Hz}}$$

14.8 SYNCHRONISING THE DISPLAY

Unless one is used to operating scope controls, a steady single waveform usually needs considerable knob twiddling to 'synchronise' the timebase to the Y-input waveform. What is happening is shown in Fig. 14.6, a typical 'unlocked' display where waveform (*a*) is traced, the spot flips back to the left hand side and then starts to sweep across at such a time that waveform (*b*) starts at a different voltage from (*a*), and then waveform (*c*) starts off at yet another different voltage. The resulting display is of several spaced waveforms, or, depending upon the frequency and your persistence of vision, the waveform 'crawls' across the screen in a worrying way.

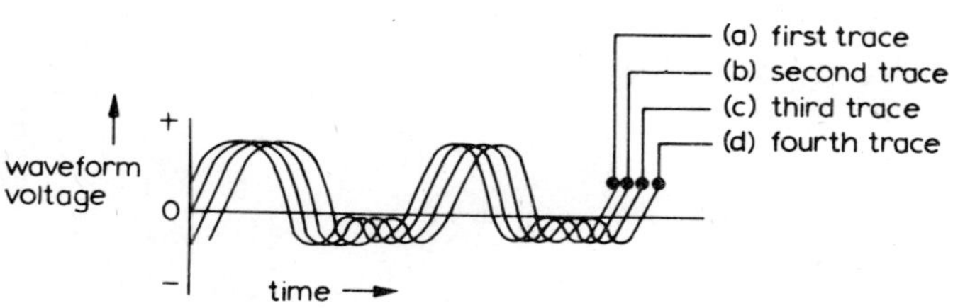

Fig. 14.6 The result of having an unsynchronised timebase

To avoid this effect, the Y waveform is also fed to the sweep generator via amplifying and shaping circuits. The 'scope' controls – usually labelled sync. (synchronising) or trig. (trigger) – then adjust the reshaped Y-waveform to 'kick' the timebase into operation at the correct instant to repeatedly trace one waveform in exactly the same place as the previous waveform. Some scopes have circuits whereby any Y-input waveform of more than about 2 mm amplitude automatically synchronises the timebase to the Y-input, producing a steady, locked display.

The X amplifier may often be switched to an external X-input terminal, allowing a graph of waveform to be drawn that does not depend upon time in the horizontal direction. For instance, the characteristic curves for many components are of voltage/current or voltage/voltage form and suitable feeds to the X and Y inputs will allow this type of curve to be drawn.

14.9 USE OF SCOPE CONTROLS

There are very few precautions to take when using the scope. In general, the knobs are there for handling and nothing can be 'blown up' by misunderstanding. An exception is the brilliance or intensity of the trace – always keep the brilliance as low as is reasonable – the trace needs to be visible, not bright enough to give you sunburn! Otherwise the main worry you will have will be the 'lost' trace, which is found in the following way:–

(*a*) disconnect X and Y inputs
(*b*) turn the brilliance well up
(*c*) set Y (or vertical) shift control to about central position
(*d*) set X (or horizontal) shift control to about central position
(*e*) turn sync. or trig. control towards 'free run',

This, with fairly small adjustments to the X and Y shift controls, should locate the trace – now turn the brilliance back to a suitable viewing level and reconnect the inputs previously removed. If the trace once more disappears it is probably that too high a voltage level (a.c. or d.c.) is being applied and the Y gain or sensitivity control needs to be turned down – usually anticlockwise – towards the 20 V/cm or 50 V/cm position to give the Y-amplifier a chance to keep the trace on the screen.

When making measurements of voltage or time, DO check that any variable gain (Y or X) or variable time (X) controls are in the 'calibrate' or cal. position, otherwise your reading will be inaccurate.

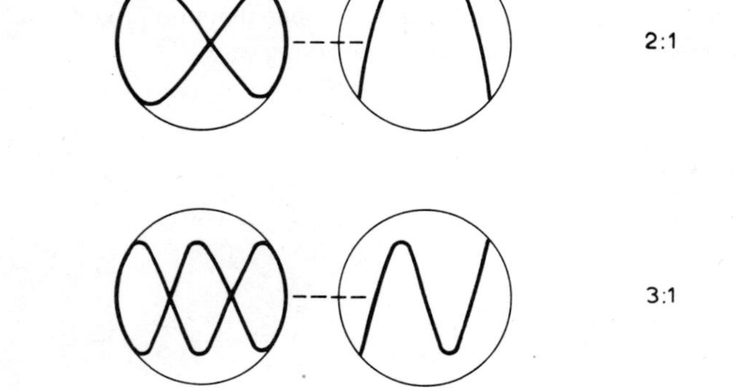

Fig. 14.7 Lissajous figures produced by feeding sine waves into both X and Y inputs

An interesting way of measuring or comparing frequencies – it makes very pretty patterns too – is to feed one sine wave to the Y-input and the other frequency sine wave to the X-input. By varying one of the frequencies changing patterns are seen. If there is a circle or ellipse drifting to a straight line the two frequencies are the same; a figure-of-eight drifting to a U or V means a 2:1 ratio, three loops drifting through an S will be 3:1 ratio, and so on. Peculiar ratios like 5:4 will give patterns that you can interpret with practice but in general remember and look for the 1:1 or 1:2 ratios. The method just described makes LISSAJOUS FIGURES and some of these are shown in Fig. 14.7. The method may be used with non-sine wave forms but the patterns then are much less easily recognisable.

Direct voltages may be measured by noting the position of the spot or trace (no timebase need be used since d.c. can produce no 'waveform' other than a horizontal line) with zero voltage input to the Y system, connecting the direct voltage to the Y-input and noting the amount of vertical shift away from the previous position. The calibration of the Y sensitivity control then allows the amplitude of the direct voltage to be calculated.

This assumes that the scope input and amplifiers (Y systems use video amplifiers) have a frequency response reaching down to zero Hertz, or d.c.; most modern scopes are switchable to measure d.c. or to include a 'd.c. blocking a.c. coupling' capacitor by means of the d.c.–a.c. switch so that a smallish a.c. waveform sitting on a high level d.c. value can have the a.c. investigated but the d.c. held back by the capacitor.

An ordinary voltmeter is usually more useful than a scope for measuring direct voltages because of the need to constantly check the position of the zero-volts deflection on the scope. With the voltmeter the value is more quickly read from the calibrated scale and there is no Y-shift control to be accidentally moved as so often happens when using a scope.

A double-beam scope may be available; one beam or display may be used by itself, but considering it as a scope with duplicated Y-inputs and Y amplifiers, it is most useful for comparing waveforms, voltage levels and phasing since the same timebase is used for both displays. When investigating power supply operation one beam, say Y_1, could be viewing the a.c. output of the transformer while Y_2 beam could be connected in turn to the rectifier cathode, the smoothing circuit and the load to see the effect of varying the different components.

Chapter 15

The generator

15.1 THE ROTATING MACHINE

The a.c. generator is an ALTERNATOR and the d.c. generator is a DYNAMO, although these terms are sometimes confused.

The simplest generator that you are likely to come across is fitted to the bicycle, where a ribbed wheel is pressed against the tyre. The faster the bicycle goes, the faster the ribbed wheel is turned and (usually) the greater the lamp brilliance. The output voltage is alternating, the result of the shaped magnet being rotated within the pole-piece wound with about a hundred turns of wire as in Fig. 15.1.

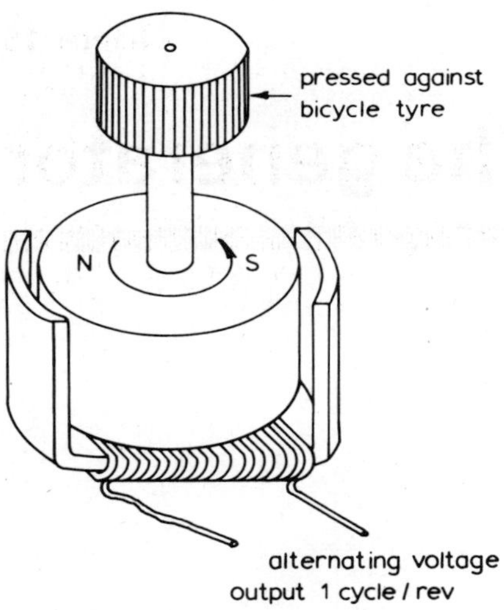

Fig. 15.1 Construction of simple dynamo or generator

When testing an amplifier, for instance to determine the frequency response or the maximum power output, whistling or speaking into a microphone does not provide a steady enough input signal and an audio generator is used. The cycle dynamo could be used to provide a signal, but, as frequencies of 1 kHz or greater are called for, the dynamo shaft would have to be rotated at 1000 rev/s or 60 000 rev/min and the whole assembly would probably fly apart! An electronic generator is needed; one able to generate or oscillate over a wide range of frequencies ald also be 'turned down' or attenuated to a low level to enable amplifier sensitivity to be tested.

15.2 THE ELECTRONIC GENERATOR

The output of an oscillator – an amplifier with frequency selective positive feedback – is liable to be loaded by the apparatus under test so that the oscillator frequency is probably shifted away from the required value by the loading. In Fig. 15.2 the oscillator is therefore followed by a buffer amplifier which isolates the load from the oscillator and provides some voltage gain. After the buffer amplifier is connected the attenuator which may feed output terminals but, rather better, is used to feed the final amplifier and then the output socket.

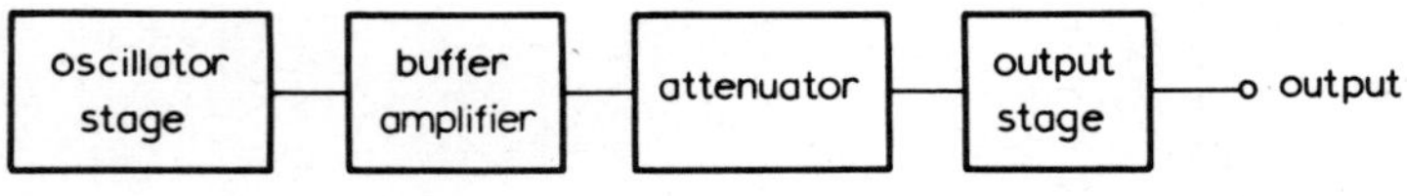

Fig. 15.2 Block diagram of electronic generator

An audio generator usually uses resistors and capacitors in the selective feedback network to determine the frequency, the common title being an *RC* oscillator. A radio frequency generator however uses coils and capacitors for frequency selection and so is termed an *LC* oscillator.

Audio generators cover the range of approximately 10 Hz to 100 kHz, r.f. generators then take over from 100 kHz to 100 MHz or so, and higher frequencies still are catered for by specialised generators covering a 2:1 or 3:1 range as from 100 – 300 MHz, 300 – 1000 MHz per apparatus.

15.3 WAVEFORMS

(i) **Sine.** The normal *RC* and *LC* oscillators are usually adjusted to generate a single frequency, in other words a sine wave only. There is a need for other waveforms however, and these are square (or pulse) and then triangular in order of popularity, with any other waveform and generator designed as needed by a particular engineer or technician.

(ii) **Square.** A simple way of generating a square wave is to over-amplify a sine wave (whose frequency is already known) to such an extent that the zero crossings from negative to positive and vice versa are so fast that the waveform at those points is nearly vertical. Over-amplification will distort the tops and bottoms of the sine wave of course, but these are clipped off to give a very good approximation to a square wave. This is shown in Fig. 15.3 where the first cycle or so is only amplified three times before clipping, the following cycle being considerably more amplified before clipping to a square wave.

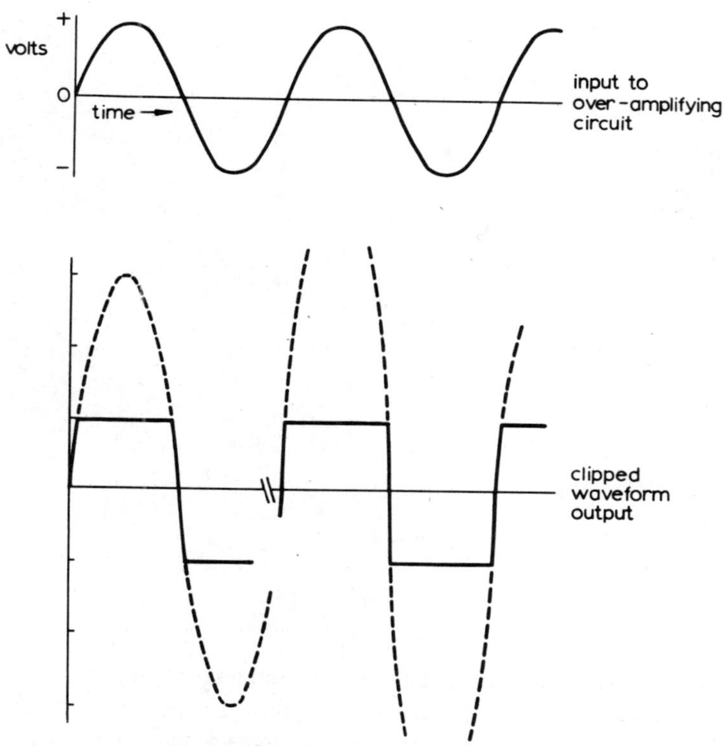

Fig. 15.3 Production of square-topped wave by overamplification of sine wave

(iii) **Ramp.** The integrating circuit has already been described in Chapter 10; a square wave fed to a resistor-capacitor coupling circuit 'loses' the leading edge of the input pulse and the output gradually rises (or falls) towards the voltage of the input as shown in Fig. 15.4. If the values of the coupling circuit are suitably chosen relative to the input-pulse frequency, the output waveform is a small-amplitude ramp, slowly rising and falling. A sawtooth waveform is a little more difficult to produce; probably the simplest circuit uses a neon, resistor and capacitor as in Fig. 15.5. The functioning of this circuit is described later on; with the values shown the neon will show a reasonable sawtooth waveform between A and B. The frequency is varied by either *C* or *R*, but be careful not to reduce *R* to less than about 100 kΩ if a small indicator neon is used, otherwise the neon will be damaged.

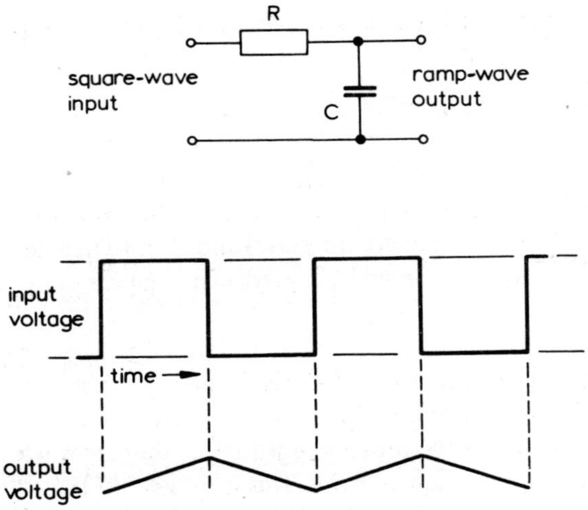

Fig. 15.4 Production of ramp waveform by integration of squarewave

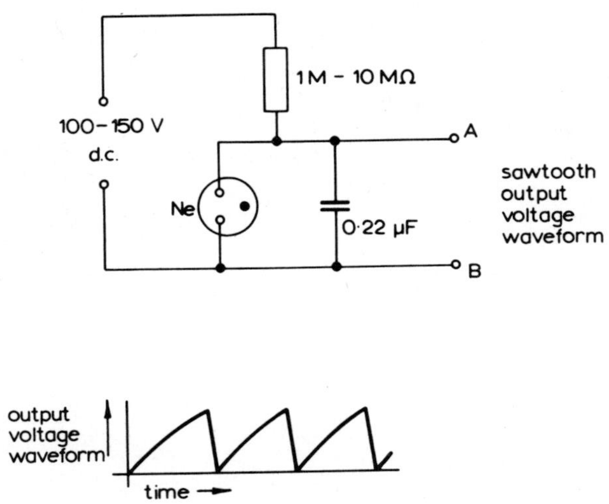

Fig. 15.5 Production of sawtooth waveform by simple neon oscillator circuit

15.4 POWER SUPPLIES

No power supplies have been shown in the block diagrams for generators but common sense says that either battery or mains operation is possible depending upon power consumption and portability. The power supplies will probably be stabilised, see Chapter 11, if the generator output frequency is also to remain stable since varying supply voltages will almost certainly cause the oscillator to change frequency slightly. This may be demonstrated (rather crudely) with the simple neon sawtooth generator by varying the d.c. supply voltage above or below the normal value and observing the change in flash rate. Also, connection of a scope across the output terminals A–B will probably change frequency, demonstrating the effect of the load upon frequency stability, and the need for a buffer between oscillator and load.

Chapter 16

Logic and gates

16.1 INTRODUCTION

The word 'logic' might be expanded into 'taking everything into consideration', so that a logical decision to sit down and to eat food demands the answers to a series of questions, such as

- am I in the mood for eating food?
- is there a chair ready to sit on?
- is the food ready for eating?
- is there time available for eating?
- are knife, fork, spoon, plate, cup etc. available?

and many other questions. The answers need not be a definite 'yes' or 'no', but the less yes and the more no an answer is, will probably require some action to correct a reply to a firm 'yes' before you carry on down the check-list and make the decision to eat.

In electronic control systems there are many instances where a logical process is used to control an output, but, unless a very complex circuit is used as in a computer, the answers to a set list of questions must be *either* yes *or* no, and there can be no correction by the apparatus to make the answers more correct.

Instead of 'answers' to questions a logical circuit looks for 'inputs'. If all the inputs are as expected – and the circuit designer decides what inputs *are* expected – the output then changes as required.

16.2 EQUIVALENT LOGIC CIRCUITS

The simplest 'yes' or 'no' equivalent is a switch which is either 'on' (closed) or 'off' (open). There is no such device as a half-closed or two-thirds-open switch; a switch can only be closed or open. Fig. 16.1 shows a very simple logical system; when the switch is closed the lamp is on, and when the switch is open the lamp is off.

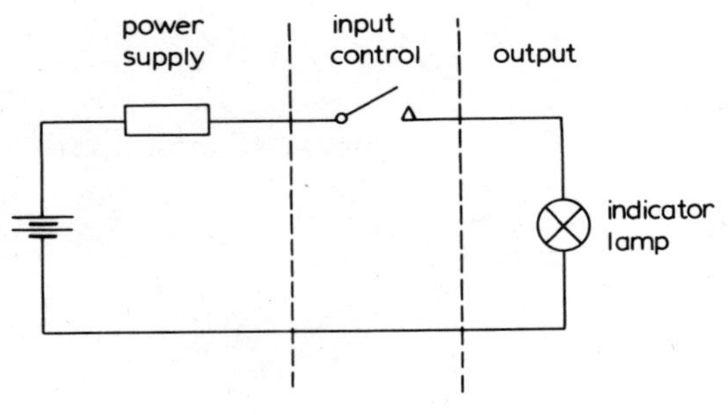

truth table

input	output
0	0
1	1

Fig. 16.1 'Input' controlling 'output' and the truth table

It soon becomes inconvenient to talk or write about inputs and outputs being yes/no, on/off, high/low, light/dark and so on, and a convention is used where 'logic–1' and 'logic–0' are used, respectively, to refer to the different states. In the subsequent equivalent circuits the following will be used:–

closed switch . . . . logical 1
open switch . . . . logical 0
on or bright lamp . . . . logical 1
off or dark lamp. . . . . logical 0

Similarly it is very longwinded to consider the various possible combinations of inputs and outputs and discuss the results in sentences, so a 'truth table' or list of input-output combinations is used instead, needing explanation only when a special set of conditions is to be considered in detail.

For Fig. 16.1, the truth table is shown below the circuit.

16.3 *AND* CIRCUIT

Fig. 16.2 shows a two-input logic circuit; switch A may be either open or closed; switch B may be either open or closed; switch B may be either open or closed; only when A AND B are closed at the same time does the lamp light. Thus there are 2 x 2 combinations of A and B inputs in the truth table.

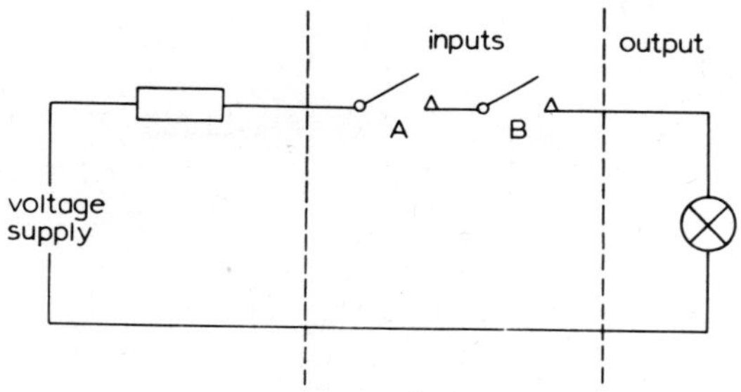

truth table

A	B	output
0	0	0
0	1	0
1	0	0
1	1	1

Fig. 16.2 2-input AND gate and truth table

In Fig. 16.3 there are three inputs; switches A, B and C. Only if ALL the switches are closed at the same time does the lamp light.

i.e. A AND B AND C light the lamp.

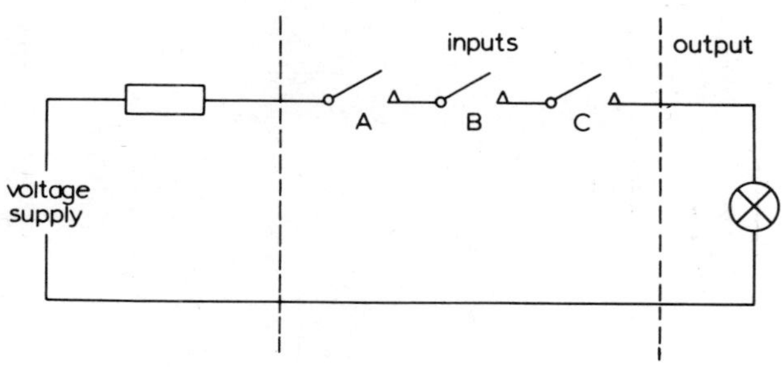

truth table

A	B	C	output
0	0	0	0
0	0	1	0
0	1	0	0
1	0	0	0
0	1	1	0
1	1	0	0
1	0	1	0
1	1	1	1

Fig. 16.3 3-input AND gate and truth table

The truth table becomes quite large because of the possible combinations of closed and open for the three switches, and is given in detail even though it is obvious (for this simple circuit) what inputs will cause the output to become logical 1.

Now although Fig. 16.2 and Fig. 16.3 show an actual circuit of an AND logical system the circuit given is only one of possibly thousands and yet all AND logical system blocks will have the same function. Rather than a square or oblong block, British Standards at present recommend the use of the symbols in Fig. 16.4 where the two-input and three-input AND-gates are given. A gate is only another name for one of the 'blocks' used in logic systems and possibly was started by an engineer thinking about the switches in series 'letting through' the signal from input to output.

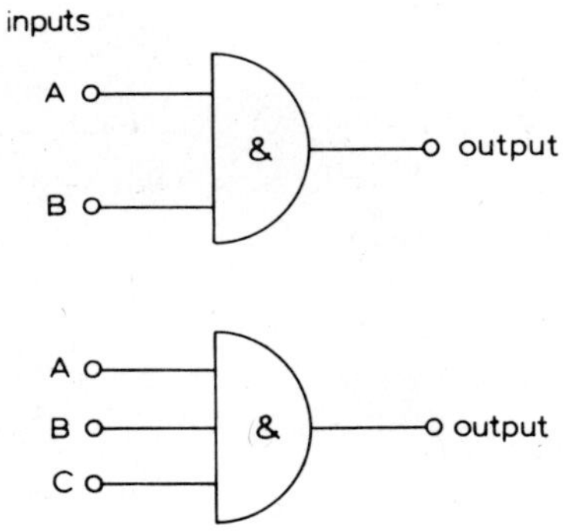

Fig. 16.4 AND gate symbols

Please disregard this section (16.4) and 16.6. They are totally confused and another source should be consulted AhD.

16.4 ~~NAND~~ *NOR* CIRCUIT

Instead of having an output produced by the presence of all logic 1 inputs it might be necessary to have the exact opposite effect – an output to be present only when all the switches are open or logical 0. The circuits need changing to that in Fig. 16.5 and it is seen that if A or B is closed the lamp is short-circuited (this is the reason for including the resistor, to avoid placing a direct short-circuit across the power supply) and only when A AND B are NOT closed is the lamp alight.

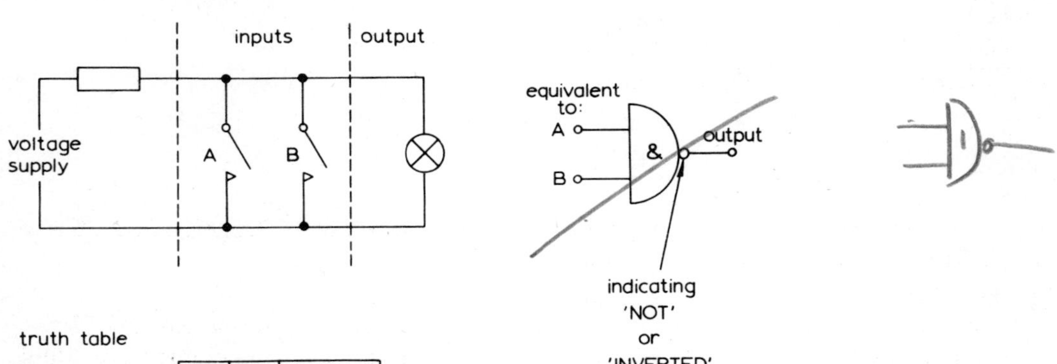

truth table

A	B	output
0	0	1
0	1	0
1	0	0
1	1	0

Fig. 16.5 2-input ~~NAND~~ *NOR* gate, truth table and gate symbol

When the inputs A = 0, B = 0 give a 1 output, a change of thinking shows that A = NOT 1, B = NOT 1 give a 1 output. Thus this gate is called a

NOT–AND type
or a BAR–AND type
or a $\overline{\text{AND}}$ type (the $\overline{}$ means 'opposite of')
or a NAND type

The word 'nand' needs careful pronunciation otherwise it is easily mistaken for 'and', the exact opposite meaning. The symbol for a NAND-gate has a small circle superimposed on the output lead, indicating the NOT applied to the function which is shown inside the gate symbol.

16.5 *OR* CIRCUIT

Another rearrangement of the switches as in Fig. 16.6 will give the result that A OR B being closed allow the lamp to light. This is thus a 2-input OR gate with the truth table and symbol as shown. The '1' inside the symbol indicates that any single one of the inputs produces a logical 1 output.

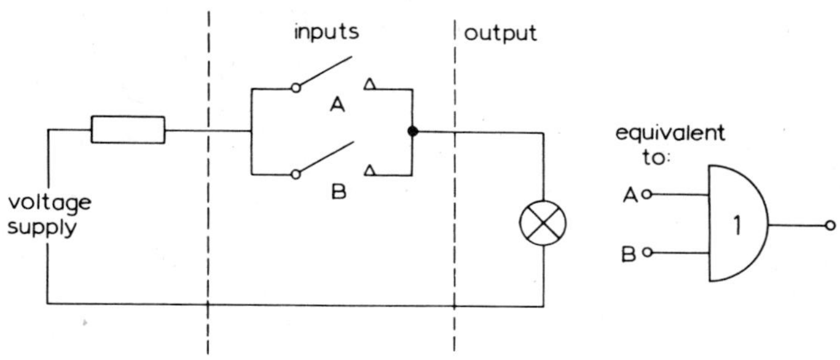

truth table

A	B	output
0	0	0
0	1	1
1	0	1
1	1	1

Fig. 16.6 2-input OR gate, truth table and gate symbol

16.6 *NOR* CIRCUIT

To produce the opposite effect of the OR circuit, the switches are arranged as in Fig. 16.7. Now, if A OR B is NOT closed, the lamp lights, as confirmed by the truth table.

A gate may have as many inputs as required – any number from one to eight, ten, twenty or as required. The truth table for a multiinput gate becomes very large but usually there are only one or two lines that matter, those giving logic–0, or logic–1 outputs, whichever are the more important in the logic system under investigation.

It is of interest to note that in the AND gate there is logic 1 output only for all-logic–1 inputs, while the OR gate gives a logic 0 output only for all-logic–0 inputs. Thus an OR gate could be said to be an AND gate for logic 0 states while the AND gate is only AND for logic 1 states. This should be kept in mind but it is better to concentrate on the following:

If 1 AND 1 AND 1 . . . gives a 1 . . . the gate is AND
1 AND 1 AND 1 . . . gives a 0 . . . the gate is NAND
1 OR 1 OR 1 . . . . . . gives a 1 . . . the gate is OR
1 OR 1 OR 1 . . . . . . gives a 0 . . . the gate is NOR

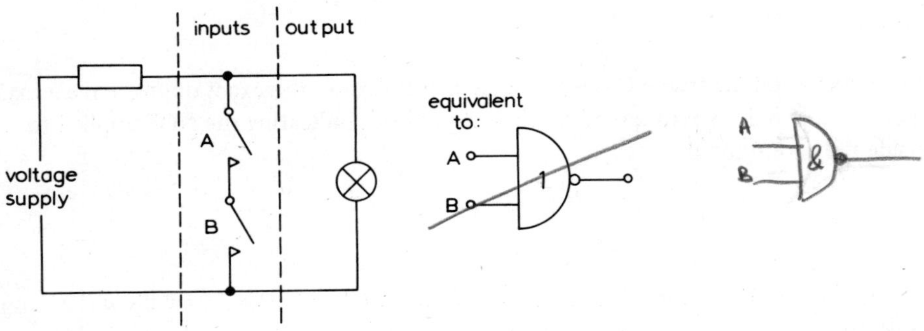

truth table

A	B	output
0	0	1
0	1	1
1	0	1
1	1	0

Fig. 16.7 2-input ~~NOR~~ NAND ~~AND~~ gate, truth table and gate symbol

16.7 *INVERT* GATE

Single-input NAND or NOR gates both share the same truth tables as in Fig. 16.8 since by referring to Figs. 16.5 and 16.7 and using switch A only, the circuits become identical. The output always becomes the opposite (or NOT) of the input and the single-input gate is thus usually called an INVERT gate or a NOT gate.

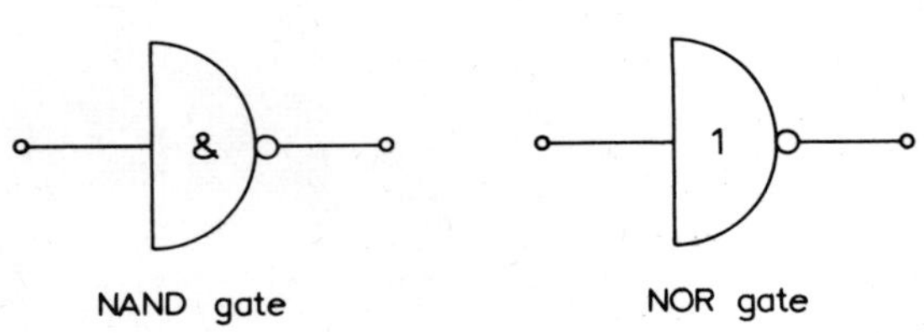

truth table

input	output
1	0
0	1

Fig. 16.8 Single-input NAND or NOR gates form INVERT or NOT gate

By adding an INVERT or NOT gate after an AND or OR gate, the pair may be replaced by a NAND or NOR. These gates are shown in Fig. 16.9. There is no significance in the use of either the AND or the OR sign within the symbol for the INVERT gate, but probably the OR symbol, 1, is the more common.

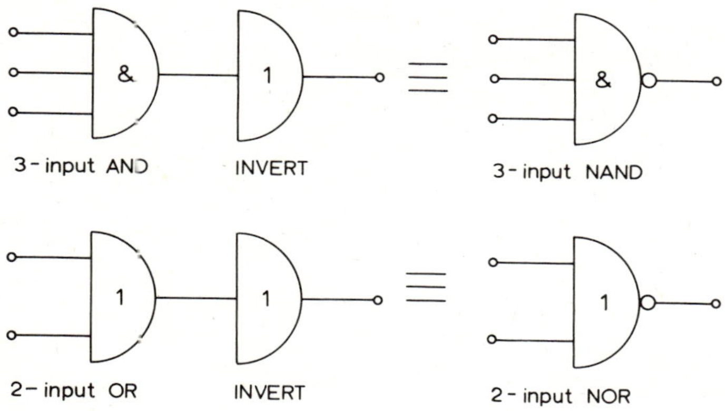

Fig. 16.9 Addition of INVERT gates to give NOT function to AND and OR gates

16.8 INPUTS

The inputs for logic circuits may be obtained from a wide variety of transducers; toggle-switches, reed switches, thermostats, fire alarm sensors, light cells, trip-wires and so on. Since a closed switch in series with a load will place a voltage across the load, a voltage input may be considered as a logic 1 input and the absence of a voltage as a logic 0 input. A very weak signal from a small sound or faint illumination may be amplified to produce a logical voltage if required, so that one of the uses of logic systems is for intruder alarms.

16.9 LOGIC ALARM SYSTEM

Consider a factory where a workroom must have traffic – people – or trolleys – allowed through, yet for reasons of safety and freedom from airborne dirt, there must be no through air-draught allowed. Doors are arranged as in Fig. 16.10*a* with microswitches or reed relays operated by the doors A, B, C, D. The contacts are closed when a door is closed, and open when a door is open. Door opening could be assisted by motors operated via photocells if necessary so that a person or object near the door results in a no-hand opening.

Now for the logical precautions:–

(*a*) A, B, C, D, never to be open simultaneously; in other words, there must always be at least one door closed.

(*b*) to allow reasonably free traffic there is a form of air-lock between A and B and between C and D; either A or B may be opened at one end of the room, and either C or D at the other end.

Without being too clever, the following logic system is suggested; see Fig. 16.10*b*. Doors A and B feed an OR gate which lights the two lamps 'wait' and 'close one door before proceeding further' both outside and inside the air-lock. Doors C and D feed a similar OR gate with two notices at C and D airlock.

Should all four doors A, B, C, D, be open simultaneously because people at each pair of doors have ignored the warning lamps and notices, the four-input AND gate then feeds the DRAUGHT alarm lamp and bell.

Another approach to the problem would be a logic system to look for a 'three-out-of-four doors open' and to then lock the remaining door until one or more of the three open doors was closed. The diagram of this is more complex than need be considered at present, although the reader might like to try drawing his own logic diagram.

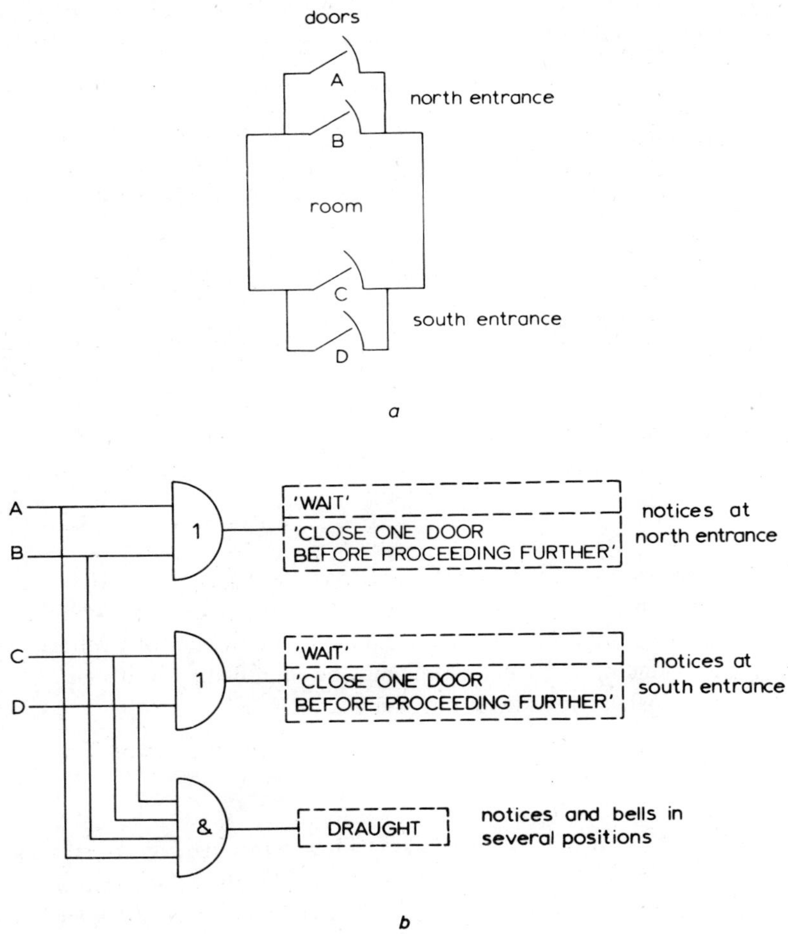

Fig. 16.10 Use of logic gates in *b* to give alarms for room in *a*

16.10 THE DIVIDER SYSTEM

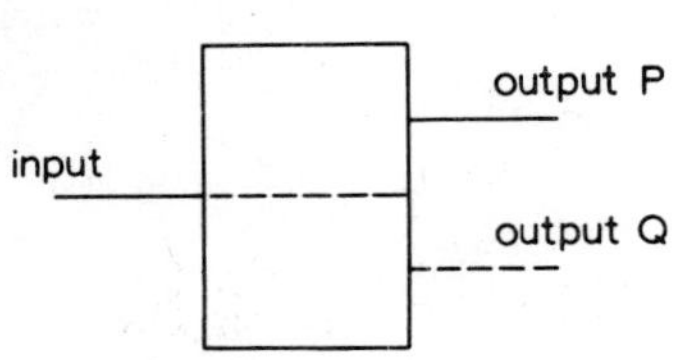

input	output P	output Q
0	0	1
1	1	0
0	1	0
1	0	1
0	0	1
1	1	0
0	1	0
1	0	1

Fig. 16.11 Bistable symbol and truth table showing divide-by-2 functioning

Another block much used in logic systems is the 'bistable', 'divide-by-two', 'flip-flop', or 'binary counter'. At its simplest this block uses two valves or transistors; in microcircuit form many semiconductor devices are used to give the required characteristics.

The symbol used is in Fig. 16.11; if the output is logical 0 to start off with, the first input pulse of logic 1, will change the output to logic 1. The second input pulse of logic 1 changes the output back to logic 0, and so on. The table for a series of inputs 0-1-0-1 shows how the output produces only half as many 1's as the input.

16.11 BINARY COUNTING

Most schools have made their pupils (or expupils) familiar with the process of counting with different 'basic' numbers, but some revision may be welcome.

We normally count in tens (probably because of our ten fingered hands) and every complete 'bundle' of ten (fingers) means another tick or unit in the tens column, spare units being numbered in the units column. Thus 4692 means (from the right-hand side)

2 odd units	or 2×10^0
9 lots of ten	or 9×10^1
6 lots of ten tens	or 6×10^2
4 lots of ten tens of ten	or 4×10^3

and note that there is no single symbol for ten, we use a combination of two symbols, 10.

Now electronic counting using only on-offs or 1—0's has only two symbols to use and so counts in twos. Thus 1011 means (from the right-hand side)

1 odd unit	or 1×2^0
1 lots of two	or 1×2^1
0 of two twos	or 0×2^2
1 lot of two twos of two	or 1×2^3

i.e. in decimal (or denary) counting

$$(1 \times 1) + (1 \times 2) + (0 \times 4) + (1 \times 8) = 11 \text{ (eleven)}$$

All the rules of binary arithmetic cannot be given in this brief chapter; to convert a decimal number to a binary number the general rule is to subtract the powers of 2 in descending order from high to low, putting a 1 in the appropriate column as you go. i.e. consider the number 147:

147 ;	256 or 2^8 is too large
128	2^7 is not too large
remainder 19	
	64 or 2^6 is too large
	32 or 2^5 is too large
less 16	or 2^4 is not too large
remainder 3	
	8 or 2^3 is too large
	4 or 2^2 is too large
less 2	2^1 is not too large
remainder 1	

	2^7	2^6	2^5	2^4	2^3	2^2	2^1	2^0
Answer:	1	0	0	1	0	0	1	1

Hence 147 (decimal or denary) = 10010011 (binary)

An alternative method is shown in the following long division by 2 process with remainders shown in the outside column.

2	147	
2	73	1
2	36	1
2	18	0
2	9	1
2	4	0
2	2	0
	1	

The binary number is now read up from the final answer through the remainder column as 1001011.

To do this type of counting a series of binary dividers is connected as in Fig. 16.12. Indicator lamps show whether an output is 1 or 0 by being ON or OFF and as pulses are fed in (to the UNITS end, usually at the right) the pulse count is then displayed in binary code.

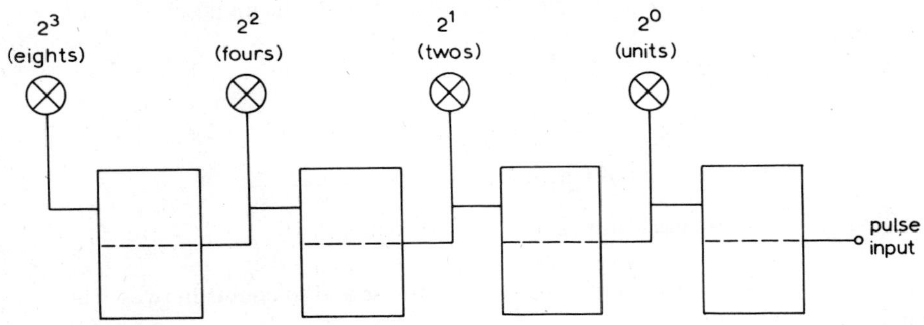

Fig. 16.12 Cascaded binary dividers

16.12 COMBINING COUNTING AND LOGIC

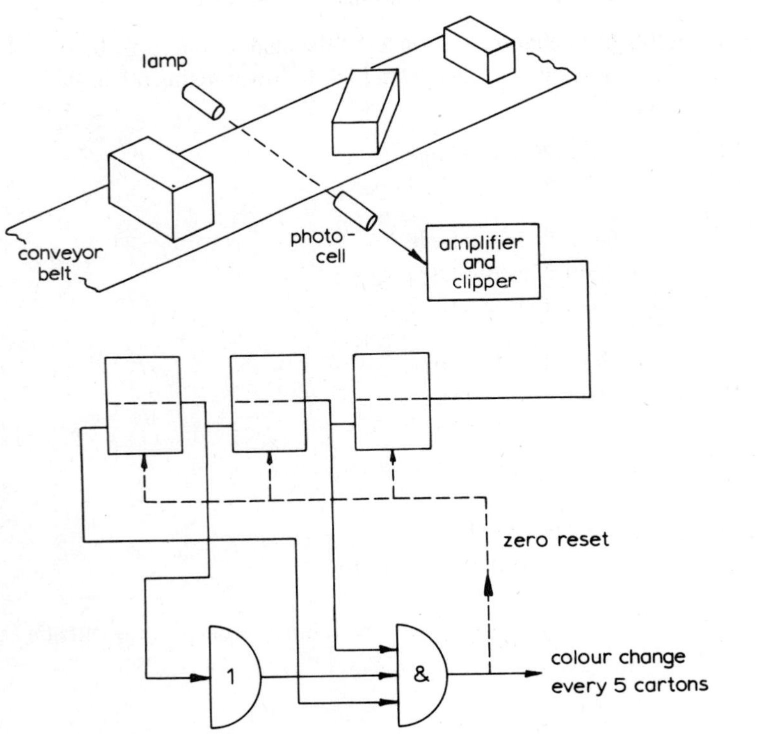

Fig. 16.13 Application of logic circuits to control of carton printing

Suppose that on a conveyor belt carrying cartons, every *fifth* carton a colour-changeover is to be operated in the printing mechanism. A photoelectric cell views the cartons, the pulses from the photocell being recorded by a binary system of the form in Fig. 16.12. By changing the 0 of a binary five's 101 into a 1 with an INVERT gate as shown in Fig. 16.13, the three 1's may be fed to a three-input AND gate so that only a binary count of 101 will produce a change-colour signal. Having counted to 101 the binary counter is immediately reset to 000 by the output from the AND gate, and the counter is then ready for the next five cartons.

Chapter 17

Television systems

17.1 INTRODUCTION

A visual scene contains such an enormous amount of information and detail that to try and send it all at once by radio is impossible. Even a simple photograph requires a complex system – camera, sensitive film, focussing and exposure, processing – to produce a print of a still picture. The eye of the viewer looking at a scene is automatically changing focus, adjusting its aperture (the iris) and looking and relooking at parts of the scene to ensure that all the wanted detail is gathered in for the brain.

There are two separate sets of information to be transmitted for a t.v. programme; vision or VIDEO and sound or AUDIO. The video, basically black and white (or monochrome) is made better by the addition of color information and even better if 3-D or stereo-vision is arranged, although this last facility is not yet available for entertainment purposes.

17.2 SCANNING THE SCENE

In 1888 a Russian inventor, Nipkow, decided upon a rotating scanning disc with small viewing holes staggered around the disc and also varying radially from the outside toward the disc centre. A scene was thus 'sliced' so that narrow strips were seen in sequence rather than all at once as is normal. In those days electronics was nonexistent and it was not until 1925 that J.L. Baird demonstrated the transmission and reception of pictures – both still and moving – by radio.

In fact, the Marconi system was adopted, since it was capable of providing a more detailed picture to the viewer. The picture was split into 405 lines (rather than Baird's proposed 30 lines) and one complete picture shown top-to-bottom 25 times per second. Recently the 625-line system has been introduced, providing even better detail together with a less 'liney' picture on large-screen cathode ray tubes. At present (1974) the 405-line system is transmitted in Britain on v.h.f. while the same programmes are transmitted on u.h.f. using the 625-line system.

17.3 THE RASTER

The black-and-white or MONOCHROME picture is a series of light or dark sections of c.r.t. screen, a scene being scanned both horizontally and vertically at the same time. A simple picture of a black triangle on a white background is shown scanned in Fig. 17.1 together with the voltage waveform of line 6. If there is no brilliance information to modulate the lines, a simple RASTER is formed on the screen, a rectangle of illumination.

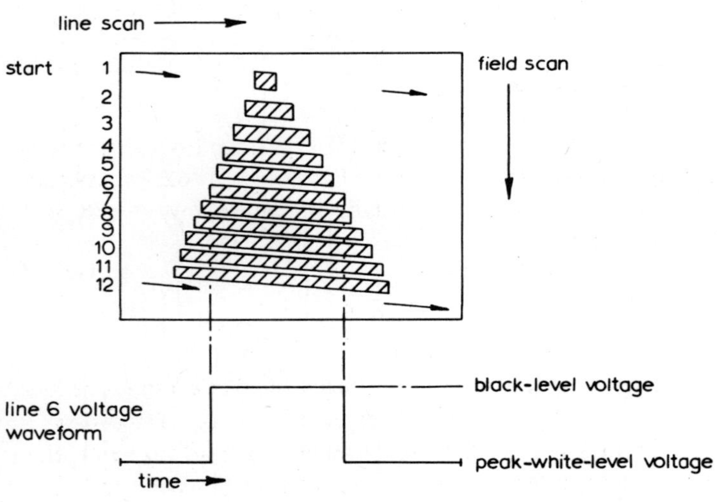

Fig. 17.1 Black triangle on white raster and waveform of line 6

In the raster the horizontal lines are produced by the LINE OSCILLATOR or LINE TIMEBASE moving the fluorescent spot of light across the screen; these lines are then slowly swept down the screen by the FIELD OSCILLATOR or FIELD TIMEBASE. One vertical scan lays down alternate lines and then the missing lines are filled in by the second vertical scan; this is called INTERLACED scanning. The 'two-field' picture gives the benefits of 25 complete pictures per second at a fairly low frequency involving some four to five Megahertz transmission bandwidth, with the advantages of 50 field scans per second that nearly synchronise to the British mains supplv and, as far as the eye of the viewer is concerned, give a smooth flicker-free picture.

17.4 THE TRANSMITTER SYSTEM

At the transmitter the scene focussed on the camera tube is being scanned – horizontally and vertically – in the same way as the raster is being produced at the receiver. To ensure that the scanning spot is at exactly the same position at exactly the same time in both camera and receiver tube, SYNCHRONISING or sync pulses are added to the video modulation from the camera to form a COMPOSITE VIDEO signal that will be used for modulation of the carrier. At the receiver the sync pulses are separated from the composite video signal and the pulses then 'kick' the two timebases – line and field – into action at just the correct instants.

Figure 17.2 shows the block diagram of a transmitter. The view seen by the camera lens and tube is sampled line-by-line, top-to-bottom by a scanning electron beam under the influence of two timebases. The sync pulses fed to the camera's timebases are also added to the amplified video signal and this composite video signal is then used to amplitude modulate the carrier for transmission.

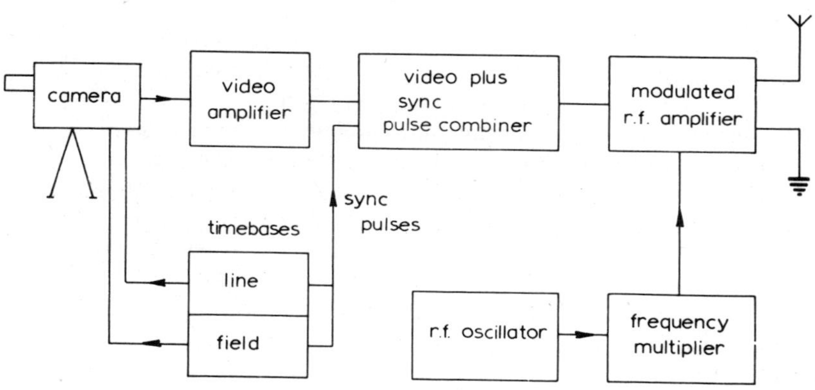

Fig. 17.2 Simple t.v. transmission system

The 405-line system uses two different a.m. transmitters for video and audio transmissions, the two carriers being 3·5 MHz apart but making use of the same aerials at both transmitter and receiver locations. The 625-line system also uses two transmitters for video and audio transmissions but this time the audio system employs frequency modulation and a carrier separation of 6 MHz from the a.m. video carrier.

17.5 RECEIVER SYSTEMS

The receiver takes the very small r.f. signal from the aerial, amplifies the signal (usually the receiver is a superhet type and frequency changing is involved) and passes it to the demodulator, where the modulation is recovered. The synchronising pulses are separated from the composite video signal, these pulses being fed off to the t.v. timebase section while the video signal is fed to the c.r.t. to control the brilliance of the spot as it scans the c.r.t. face.

Fig. 17.3 shows a much simplified block diagram of a 625-line receiver – it must be realised that there are many more ways of drawing this diagram depending upon the detail required.

A u.h.f. modulated carrier (which you will not see on a scope unless you have access to 1000 MHz type!) uses modulation whereby the blacker the scene viewed, the higher the r.f. carrier level transmitted. This is the opposite of the v.h.f. modulation method but at the demodulator the only change needed is reversal of the demodulator diode to produce a normal type of picture. You are probably familiar with special effects on t.v. where a picture is made to 'go negative' – usually science fiction plays – when the programme producer introduces a phase-inverting amplifier in the modulation chain.

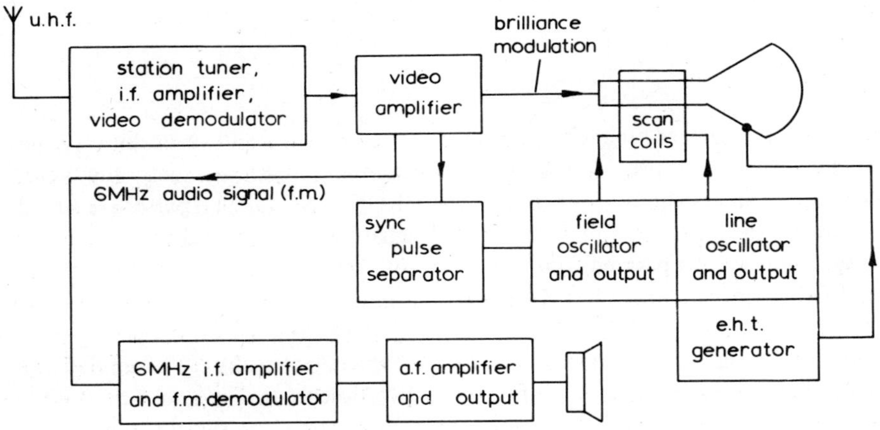

Fig. 17.3 Simple monochrome t.v. receiver system

17.6 BANDWIDTH REQUIREMENTS

The fast change from black to white in 405-line detailed scenes requires video modulation frequencies of up to 3·5 MHz and this would involve an a.m. channel 7 MHz wide with normal upper and lower sidebands. To save frequency space and allow closer channel spacing the upper sideband is largely removed by suitable filtering so that the actual spectrum used is shown in Fig. 17.4, each complete t.v. channel having 5 MHz spacing.

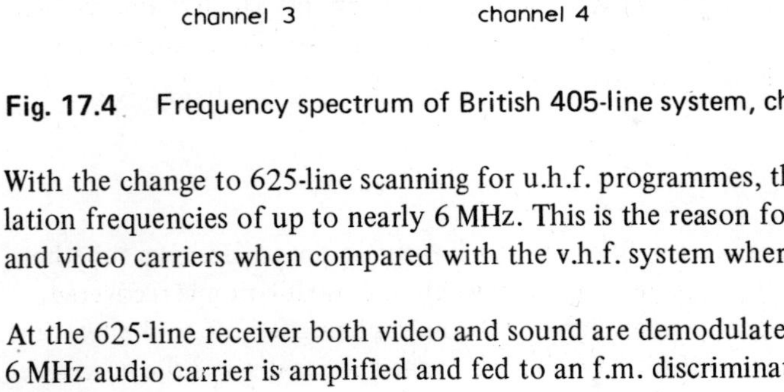

Fig. 17.4 Frequency spectrum of British 405-line system, channels three and four

With the change to 625-line scanning for u.h.f. programmes, the greater detail now possible involves video modulation frequencies of up to nearly 6 MHz. This is the reason for the increased frequency spacing between the audio and video carriers when compared with the v.h.f. system where 3·5 MHz spacing is used.

At the 625-line receiver both video and sound are demodulated as a composite wide-band signal and then the 6 MHz audio carrier is amplified and fed to an f.m. discriminator (or demodulator) for the actual audio signal to be recovered.

The channel spacing on u.h.f. is 8 MHz, channels 21–34 being in the 470–582 MHz band and channels 39–68 being in the 614–854 MHz band. Having a large frequency difference between the low and the high channels for one district means that aerials have to be made-to-measure for groups of stations. Each area usually is allocated four stations in an 88 MHz bandwidth, i.e. channels 53–63, or 23–33 or 41–51, thus allowing ample separation between actual tuning points in a receiver but still enabling a single aerial to be used for the area stations.

17.7 COLOUR

In addition to the brilliance – which now is called LUMINANCE – and line and frame synchronising signals to be conveyed, there is also the colour – or CHROMINANCE – and colour synchronising information to be included in the modulation. The system is considerably more complex and is dealt with only briefly.

At the transmitter a 'colour-burst' of a few cycles of 4·43 MHz is inserted during the period just before the left-hand side of a picture line starts across the screen. At the receiver this 'burst' is used to correct and maintain a local oscillator within the receiver at exactly the correct frequency for colour synchronising purposes at the colour demodulator. The colour information is then decoded and used to control the emission of the three electron guns within the colour-tube.

The colour tubes at present used have three beams of electrons steered very accurately through holes or slits to collide only with their own dot or stripe of phosphor. One electron gun produces only a red fluorescence, another gun produces green and the last gun produces blue. All three colours – RED, GREEN and BLUE – produced in just the correct quantity, add to give the effect of white to the viewer while different proportions of the three colours will produce the other missing colours.

Chapter 18

Other systems

18.1 RADAR SYSTEMS

In a large hall, a single person whistling or making a sharp noise expects to hear an echo. With a little practice the size of the hall may be judged by the timing and the amount of echoes and this method is in fact used by blind people. Radio energy travels far faster than sound energy – 3×10^8 m/s compared with about 330 m/s – and in about 1935 electronic components had been developed sufficiently to allow the length of time between the transmission of an r.f. pulse and its reception only a few microseconds later to be accurately measured.

The resulting technique was called RADAR, a term now generally used to cover the measurement of range, height, speed, bearing and often size of an object.

The directional aerial in Fig. 18.1 is switched to the pulse transmitter and a pulse of r.f. is sent off into space. The aerial is immediately switched to the receiver and the very small amount of reflected energy is amplified, detected and displayed on the c.r.t. The trace on the c.r.t. is moved by the timebase whose speed is accurately known or calibrated, and the S and R blips on the screen (caused by the r.f. sent and received) thus have an accurately known time interval between them. *Half* of this time is the time taken for the r.f. to travel – at 3×10^8 m/s – from the radar unit to the reflecting aircraft, hence the distance to the aircraft is now known.

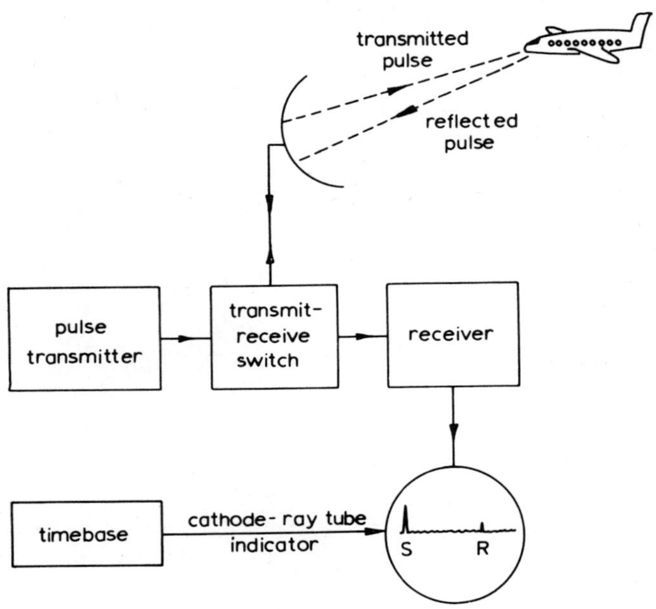

Fig. 18.1 Basic radar system used to determine range to aircraft

If the aerial is large and precisely made, the bearing and elevation are derived from the mechanism that steers the aerial looking for reflected pulses, while aircraft speed is computed from the rate-of-change of distance.

18.2 DOPPLER RADAR

A radar motor vehicle 'speed-trap' usually uses a slightly different principle from that already described. You are probably familiar with the way in which the horn (or whistle or bell) of an approaching vehicle sounds higher in frequency than when the vehicle is departing. This may be explained briefly by imagining the emitted sound waves from the vehicle at the *actual* frequency being 'bunched' at the listener because the vehicle is bringing the sound source towards the listener who hears a *higher* frequency. When departing, the waves from the sound source are 'stretched' as far as the listener is concerned, and the frequency appears to be *below* that of the source.

Radio waves react in a similar way; a beam of r.f. energy is transmitted by the aerial in the speed-trap box at the side of the road and the reflected r.f. is then checked closely for the frequency difference between the transmitted and received frequencies. The larger the difference, the more the read-out instrument deflects and the faster the approaching car. It is no good moaning about the unfairness of it all; electronics is our trade and we all have our own speed detectors – eyes and speedometer!

18.3 POSITION FINDING

While radar systems are usually understood to be a combined transmitter-receiver set-up, the position finding apparatus carried by many ships and aircraft involves the use of fixed land-based transmitting stations. At least three stations transmit a modulated r.f. signal, the modulation being exactly in phase (relative to time) at the instant of transmission.

The distant receiving station of the ship (or aircraft) receives two signals, and depending upon the distance of each transmitter from the ship, calculates from the difference in modulation phasing the time separation between the two signals arrival. In Fig. 18.2, no time difference between signals from transmitters X and Y places the ship somewhere on the line *PQ*. If signal X is early compared with signal Y, the craft is on the X side of *PQ*; if signal Y is the earlier, the craft is on the Y side of *PQ*.

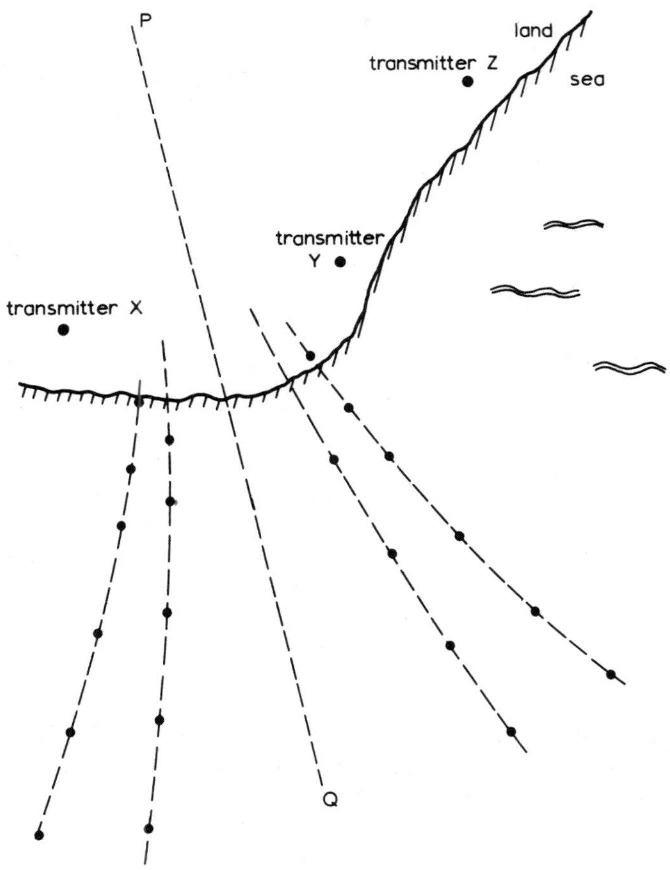

Fig. 18.2 Plan on which signal time-delay allows position of craft to be determined

Charts overprinted with dotted lines for various difference-in-arrival times are issued by the equipment manufacturers for the area in which the craft is operating; the craft is thus located somewhere on one of the dotted lines according to the signals from the X-Y transmitters.

The receiver is now returned to select, say transmitters X and Z, the arrival-of-signal phase difference noted once more and a different dotted line on the chart found. Since the craft could only be on *one* line for the X-Y transmitters and is then found to be on *another* line for the X-Y transmitters, the intersection of the two lines is the position of the craft.

18.4 ANALOGUE–TO–DIGITAL CONVERSION

One method of achieving this is to convert the analogue quantity (metres, height, speed, gallons, current etc.) from its ordinary measurement units into voltage via some form of transducer. This value of voltage is then used to control the frequency of an oscillator whose frequency is counted in binary form and displayed or stored as a series of 1's and 0's. Fig. 18.3 suggests a method whereby the quantity of liquid in a vessel has the height (hence quantity) transmitted by a float to a variable resistor. The voltage picked off by the slider controls the oscillator frequency, but it is no good just counting continuously the pulses from the oscillator, the indicated quantity

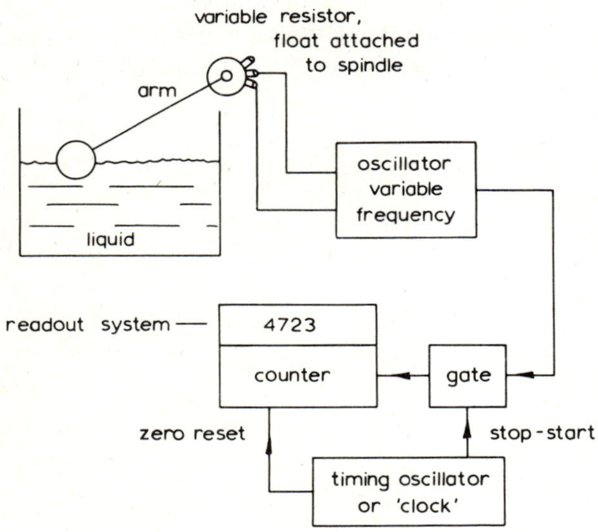

Fig. 18.3 Liquid height or quantity displayed in digital form

keeps on and on increasing! A separate system is needed to start the count, stop the count and display for a while, reset the count to zero and then start all over again; this subsidiary timer or clock is shown in Fig. 18.3 in simplified form. It will be noticed that the read-out display is in fact giving a decimal (or denary) read-out, not a binary display, so yet another system – a binary-to-decimal convertor – is involved.

18.5 BINARY-TO-DECIMAL CONVERSION

As already stated, counting is most easily done electronically by using the binary code and logic gates, but a disadvantage is that the normal person has been taught to recognise the numbers 0–9 and he (or she) takes some time to work out the meaning of 1000, for instance. And then changing from a machine where a readout of 1000 means 8 to another machine where 1000 this time means a thousand leads to difficulties!

The binary numbers for denary 0–9 are all different, as shown in Fig. 18.4*a* and only one number in the range 0–9 must be illuminated or driven for each binary combination. For instance *one* could be selected by checking (electronically) to see if the first three 0's of 0001 are there – but they are also present for a *nought*. So it would be better to check for the first three 0's being present AND the last 1 as well – or perhaps invert the first three 0's into 1's and feed the resulting binary 1111 into a 4-input AND gate, the output of this gate then driving the *one* figure in the denary code.

Similarly *two* could also have its binary 0's inverted into 1's and another 4-input AND gate used to drive the two figure in denary code. The overall logic circuit is not too complicated but it is such a common code conversion problem that the logic manufacturers produce integrated circuits with all the interconnections already made. Only '1–2–4–8' or 'A–B–C–D' inputs are needed and ten outputs are available to drive the denary read-out figures. Fig. 18.4 *b* shows the start of a B–D convertor, as it is often termed, to give denary *one* and *two* outputs.

binary D	binary C	binary B	binary A	denary
0	0	0	0	nought
0	0	0	1	one
0	0	1	0	two
0	0	1	1	three
0	1	0	0	four
0	1	0	1	five
0	1	1	0	six
0	1	1	1	seven
1	0	0	0	eight
1	0	0	1	nine

a

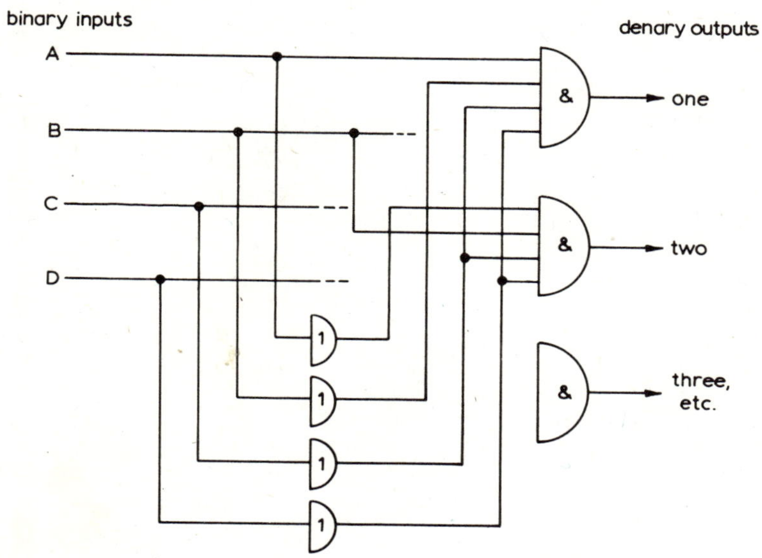

b

Fig. 18.4 Table of binary and denary equivalents with part of the logic binary-to-decimal conversion diagram

18.6 THE DIGITAL COMPUTER AND APPLICATIONS

The digital computer does arithmetic and calculations – fast and accurately – but only as well as instructed by the computer programmer via the input system. The 'language' may be called Fortran, Cobol or other titles, depending upon the type of computer used and the way in which the computer is told to 'add' 34 to 17; square the result; take this away from the stored figure of five calculations back; see if the result is now larger or smaller than 56; . . . . The computer thus needs to contain a storage system and an arithmetic system, finally feeding the results out via the output system on punched tape or printed paper.

In electronics one application is for computer-aided design – or c.a.d. – where a new design or circuit is analysed, the values of the components being converted into digital or binary form and, via the input, stored in the memory. The computer is now instructed or programmed to compute the many values of voltage, current, power dissipation, frequency and so on present at each component. The programme may also call for the computer to try changing the value of one component at a time to the expected variation of ten percent high – recompute all values – then ten percent low – recompute all values – change next component ten percent high . . . and then if necessary to try the effects of having two components high in value, then three components and so on.

A vast amount of information is available from this work, the calculations being done at a possible rate of 10 000 per second, the results being probably stored temporarily in the memory system. The programmer may then instruct the computer to search all this data for the best (or safest, or least distorted) conditions of operation of the circuit and to print out the component values giving the required condition. As a bonus, the printout could also indicate those components which may be low tolerance for cheapness and those which must be more accurate. And – if the information is already stored in another memory – there could also be a printout of the component manufacturer, price, discount and availability as well!

A computer and memory stores like these costs many thousands of pounds and the running costs are not small; c.a.d. is only economical for a circuit that will be made in quantity, but a microcircuit that is successful may be used in millions and only cost a few pence when made.

Computers and their associated stores are also used for wage calculation, store keeping, calculation and issuing of invoices (gas, electricity, telephone, etc.) and are usually the basis of the pocket calculator.

Index